Alphonso Wood

How to Study Plants

Or introduction to botany Being an Illustrated Flora

Alphonso Wood

How to Study Plants
Or introduction to botany Being an Illustrated Flora

ISBN/EAN: 9783337106669

Printed in Europe, USA, Canada, Australia, Japan

Cover: Foto ©berggeist007 / pixelio.de

More available books at **www.hansebooks.com**

CYATHEA ARBOREA,—A TREE FERN.

Scene on the Chagres River, Central America.

HOW TO STUDY PLANTS,

OR,

INTRODUCTION TO

BOTANY,

BEING

AN ILLUSTRATED FLORA.

BY

ALPHONSO WOOD, A. M., Ph. D.,

AUTHOR OF "THE CLASS-BOOK OF BOTANY," "OBJECT LESSONS IN BOTANY,"
"PLANT RECORD," ETC.

AND

Edited by J. DORMAN STEELE, Ph. D., to accompany the
"FOURTEEN WEEKS SERIES IN NATURAL SCIENCE."

*"There breathes, for those who understand,
A voice from every flower and tree;
And in the work of Nature's hand
Lies Nature's best Philosophy."*

COPYRIGHT, 1879, 1882, BY

A. S. BARNES & COMPANY,

PUBLISHERS,

NEW YORK AND CHICAGO.

WOOD'S BOTANY.

OBJECT LESSONS IN BOTANY, pp. 340, 12mo. An introduction to the Science, full of lively description and truthful illustrations; with a limited Flora, but a *complete* System of Analysis.

THE BOTANIST AND FLORIST, pp. 620, 12mo. A thorough textbook, comprehensive and practical; with a Flora, and System of Analysis equally *complete*. "I have been deeply impressed, almost astonished, (writes Prof. A. Winchell, of University of Michigan) at the evidence which this work bears of skilful and experienced authorship—nice and constant adaptation to the wants and conveniences of students in Botany," etc.

THE CLASS-BOOK OF BOTANY, pp. 850, 8vo. The principles of the Science more fully announced and illustrated—the Flora and Analysis complete with all our plants portrayed in language, both scientific and popular. "The whole science (writes Prof. G. H. Perkins of Vermont University), so far as it can be taught in a college course, is well presented, and rendered unusually easy of comprehension. I regard the work as most admirable."

THE PLANT RECORD—a beautiful book, for classes and amateurs, showing, in a few pages, how to analyze a plant—any plant, and furnishing tablets for the systematic record of the analysis.

FLORA ATLANTICA, or **WOOD'S DESCRIPTIVE FLORA**, pp. 448, 12mo. This work is equivalent to the Part IV of the Botanist and Florist, being a succinct account of all the plants growing East of the Mississippi River, both native and cultivated, with a system of analytical tables wellnigh perfect.

WOOD'S BOTANICAL APPARATUS—a complete outfit, for the field and the herbarium. It consists of a portable trunk, a Wire Drying Press, a Knife-trowel, a Microscope, and Forceps.

"FOURTEEN WEEKS" IN EACH SCIENCE,

By J. Dorman Steele, Ph. D., F. G. S., Etc.

Now Ready:

PHILOSOPHY.	PHYSIOLOGY.	ZOOLOGY.
CHEMISTRY.	GEOLOGY.	ASTRONOMY.

A KEY to *Practical Questions in Steele's Works,*
Seven volumes, each, $1.00.

STEELE'S BOT.

PREFACE.

THE plan of this work differs from that of the ordinary Botany. The method pursued is to introduce the pupil at once to the study of the plant itself, by means of elaborate illustrations, and living specimens.

The parts and functions, together with the generic and specific characters of each plant, are pointed out and described. The thing being seen, is then named. No new term is introduced until a necessity arises for its use. About one hundred representative plants are thus explained. The work may therefore be considered as a limited Flora. But it is much more. Through an acquaintance with these representative plants, the pupil is gradually led to a knowledge of the principles of Botany. In the common treatise he is told the general law and then given illustrations; in this, he is shown the instances, and thence conducted to the broad truth of Nature.

The selection of plants for analysis has been determined by the following considerations: (1.) The plant should be common throughout the country and hence accessible to every learner; (2.) It should flower in the spring or early summer, that being the season when the study is generally pursued in our schools: (3.) It should have conspicuous parts, at least the earlier ones, adapted to the comprehension of a beginner; (4.) It should belong to one of the more important Orders, as neither the limits of the book nor the requirements of the plan adopted would admit the study of them all; finally, the selection was often influenced by some intrinsic feature which fitted the plant to illustrate a special principle in vegetable life, as sleep, irritability, cross-fertilization.

This work is merely an introduction, conducting the pupil across the gateway only. Yet it is not designed for infants; the rather for learners capable of thought and reason. To all such it offers a helping hand, seeking to smooth their path and to awaken such an interest in the subject as will induce them to pursue their investigations in more advanced books and in the broader field of Nature itself.

The illustrations in this work are nearly all from original designs prepared by Mr. Sprague, "the most accurate of living botanical artists," and Mr. Emerton, designer of the illustrations in Prof. Eaton's North American Ferns.

SUGGESTIONS TO TEACHERS.

WHENEVER possible the plant described in the lesson should be in the pupil's hand for examination. This is the very life of the recitation. A constant supply of specimens should therefore be secured for this purpose. Let them be sought a day in advance and in the order of the book. The arrangement may, however, be changed when necessary to accommodate the collector's convenience. Should any terms then arise not already explained, their meaning can always be found by reference to the Glossary at the close of the book. After a few lessons have been thoroughly understood, the pupil need no longer confine his attention to the few plants treated in the text. He can readily repeat the process of analysis on any specimen he is able to secure. He should, however, be admonished that this will lead to desultory habits of study unless he completes every analysis which he begins, and records the result, in his ever-present memorandum book. The strictest care should be observed in completing the Tablets of this work or of the Plant Record. They

should be carefully and neatly filled up from notes previously arranged, and adjusted in every word and sentence, so that there may be no erasures and no interlining, and the Record may represent in every particular the pupil's best work. Blank forms should be drawn on the blackboard at every recitation, and pupils be required to complete them, subject to the criticism of the teacher and of the class as to analysis, expression, style, spelling, punctuation, etc.

A microscope is essential to botanical work. Small hand-magnifiers for the use of the pupils and a larger table-instrument for the teacher, can be procured of the publishers of this book, Messrs. A. S. Barnes & Co., 111 and 113 William St., New York. The *Flora Atlantica*, or Wood's Descriptive Flora, is the proper sequel or companion of this treatise. By means of an elaborate system of analytical tables, the student, with a plant in hand, is unerringly guided to its name, classification and history. For this interesting work he is thoroughly prepared by the lessons in this treatise. For collecting specimens and drying them for the herbarium, there are required, (1) a tin box or trunk shutting closely; (2) a drying press of woven wire and bibulous paper; (3) a knife-trowel. They may be obtained of the publishers above mentioned. A system of questions for study or review, generally applicable to all plants, will be found in the Appendix. It is recommended that an herbarium including, at least, all the species described in this work, be provided for use in class exercises in the absence of any fresh specimen.

TABLE OF CONTENTS.

CHAP.		PAGE
I.	PIGEON-WHEAT MOSS—Polytrichum. The Analysis. Capsule. Operculum. Calyptra. Peristome. Spores. Flowers. The Flowerless Plants. How the Moss grows.	13
II.	THE APPLE MOSS—Bartramia. The double peristome. Fugacious calyptra. Cellular structure. THE MUSCI.	17
III.	THE POLYPOD FERN—Polypodium. The rhizome. Frond. Sporangia. Spores. Forked venation. How the Fern grows.	20
IV.	THE OSMUND FERNS—Osmunda. The vernation. The species. THE ORDER FELICES. Tree Ferns. THE CRYPTOGAMIA. The uses of Ferns. The Climbing Fern. The Brake. The pioneer vegetation............	24
V.	THE DOGTOOTH VIOLET—Erythronium. The two Regions. The bulb. Leaf. Venation. The Calyx. Corolla. Stamens. Pistils. The fruit. Seeds. Pollen. The province ENDOGENS. THE PHENOGAMIA....................	29
VI.	THE TULIP—Tulipa. The tunicated bulb; its contents. The flower. Varieties. The Tulip mania in Holland....	35
VII.	THE SPRING BEAUTY—Claytonia. Tubers. A raceme. The petals and their colored lines. Opposing stamens. The seed and its albumen. Æstivation. Our two species. The PORTULACACEÆ. The Portulacas. The province EXOGENS.	39
VIII.	THE EARLY CROWFOOT—Ranunculus. Fasciculate roots. Perennial herbs. The nectary. Polyandrous and hypogynous stamens. The simple fruit—distinct carpels.	46
IX.	BULBOUS CROWFOOT—Ranunculus. An inaxial root. The corm. Reflexed sepals; economy. Plan of the flower..	50
X.	THE LIVERLEAF—Hepatica. Crown-stem. Palmate venation. Involucre. Apetalous flowers. Anatropous seeds.	54
XI.	RUE ANÉMONE—Anemône. Tuberous root. Umbel. Compound leaves. Distinctness of organs. Absence of honey.	58
XII.	WOOD ANÉMONE—Anemône. Creeping root; rhizome. Solitary inflorescence. Species. Order RANUNCULACEÆ.	60
XIII.	BLOODROOT—Sanguinaria. Rhizome. Juice. Caducous sepals. Parietal placentæ. Dicotyledonous embryo.....	64
XIV.	THE POPPY—Papaver. An annual herb. The species. Order PAPAVERACEÆ. The California Poppy. Celandine. Use and culture of Opium......................	68
XV.	THE VIOLETS—Viola. A cucullate leaf. Resupinate, irregular flowers. Adnate anthers. Cleistogene flowers. Economy in pollen........................·...........	71

CHAP.		PAGE
XVI.	THE GARDEN VIOLET—Viola tricolor. Lyrate-pinnatifid stipules. Auriculate sepals. Order VIOLACEÆ. Species.	75
XVII.	CHICKWEED—Stellaria. Nodes. Internodes. Centrifugal inflorescence. Bifid petals. Free central placenta.	78
XVIII.	THE PINK—Dianthus. Caudex. Caulis. The Calyx as a flower-cup. Proterandrous flowers. Teratology. Order CARYOPHYLLACEÆ	80
XIX.	THE WILD GERANIUM. Nodes. Internodes. Stipules. Regma. Carpophore. Folded cotyledons. Herb Robert.	83
XX.	THE HORSE-SHOE GERANIUM—Pelargonium. The Spur.	86
XXI.	YELLOW WOOD SORREL—Oxalis. Leaf trifoliate. Leaflet obcordate. The leaf-axils Monadelphous stamens. Contorted æstivation. Sleep of plants. The Order....	89
XXII.	JEWEL WEED—Impatiens. Corolla irregular and spurred. Irritable fruit. Contrivances for scattering seeds....	93
XXIII.	NASTURTIAN—Tropæolum. Peltate leaves. Spurred sepal. Unguiculate petals. The order GERANIACEÆ.	95
XXIV.	SHEPHERD'S PURSE—Capsella. Leaves amplexicaul. Flowers cruciform, tetradynamous. A silicle.......	89
XXV.	THE TOOTHROOT CRESS—Cardamine. A silique. Cotyledons accumbent—incumbent. Order CRUCIFERÆ...	101
XXVI.	STRAWBERRY—Fragaria. Scape. Cyme. Perigynous stamens. Strawberry fruit. Quincuncial æstivation. Hairs.	104
XXVII.	THE APPLE TREE—Pyrus. Trunk. Wood. Medullary rays. Annual layers. Food of plants. Circulation of the sap. Ovary adherent. Fruit a pome. Seed. Germination.	107
XXVIII.	THE ROSE—Rosa. History. The Prickles. Odd-pinnate leaves. Ovary inferior. Seed suspended. The Hip. The Double Rose. The order ROSACEÆ. Peach, Quince, Blackberry, Spirea, etc..................	112
XXIX.	THE PEA—Pisum. Tendrils. their action. Papilionaceous flower. Diadelphous stamens. Legume......	117
XXX.	THE LOCUST TREE—Robinia. Stipular spines. Sensitiveness. The Sensitive Plant. The Moving Plant. The order LEGUMINOSÆ.................	121
XXXI.	THE EVENING PRIMROSE—Œnothera. Leaves spirally arranged. Root biennial. Calyx adherent, tubular. Flowers nocturnal................	125
XXXII.	LADY'S EARDROPS—Fuchsia. Angular pollen grains. Hybridization. Order ONAGRACEÆ. Zauschneria...	128
XXXIII.	SWEET CICELY—Osmorhiza. The axial root. Decompound leaves. Sheathing petioles. Compound umbel. Involucels. The cremocarp. Carpophore......	131
XXXIV.	GOLDEN ALEXANDERS—Carum. Ovary inferior. Ribs and vitæ of the fruit. Oil tubes. Action of light. The order UMBELLIFERÆ................	134

CONTENTS.

CHAP.		PAGE
XXXV.	THE MOUSE-EAR EVERLASTING—Antennaria. Stolons. Diœcious plants. Heads of florets. Receptacle. Involucre. Pappus clubby. Cypsela	137
XXXVI.	ROBIN'S PLANTAIN—Erigeron. Heads radiate. Florets of the ray. Florets of the disk. Ligulate corolla	140
XXXVII.	THE DANDELION—Taraxacum. Runcinate leaves. Radiant, homogamous heads. Chaff. Syngenecious anthers. The order COMPOSITÆ. Chickory, Camomile, Aster. Chrysanthemum. Solidago	143
XXXVIII.	THE CHECKERBERRY—Gaultheria. Urceolate corolla. Curious fruit. The Black Checkerberry	147
XXXIX.	THE PYROLAS. Anthers inverted in bud; opening by pores. The six species	150
XL.	PRINCE'S PINE—Chimaphila. Horned anthers	152
XLI.	THE KALMIAS. Elastic stamens. Pollenization. The order ERICACEÆ. The Heaths. Blueberries. Cranberries. Azalias. May-flower	155
XLII.	THE PITCHER PLANT—Sarracenia. Ascidia. Order SARRACENIACEÆ. Carnivorous Plants. Venus' Flytrap	158
XLIII.	THE AMERICAN COWSLIP—Dodecatheon. Opposing stamens. Dimorphism. Free central placenta	161
XLIV.	CHICK WINTERGREEN—Trientalis. 7-parted flowers	164
XLV.	THE LOOSESTRIFES—Lysimachia. Verticillate leaves. Monadelphous stamens. Opposing stamens explained. Order PRIMULACEÆ. Cyclamen. Anagallis. Primrose.	165
XLVI.	THE SPEEDWELLS—Veronica. Why so called. A two-celled capsule. Exserted stamens. The species	170
XLVII.	TOAD FLAX—Linaria. Pentamerous flowers. Personate corolla. The spur,—what can reach its honey. Order SCROPHULACEÆ. Digitalis. Pentstemon	173
XLVIII.	THE GROUND IVY—Nepeta. Naturalized plants. Bilabiate corolla. Halved anthers. Seeds apparently naked. The Catmint	176
XLIX.	BLUE CURLS—Brunella. Cuspidate bracts. Hairs jointed. The lip a doorstep for bees. Order LABIATÆ. Peppermint. Oil of Spike. Lavender	179
L.	THE MORNING GLORY—Ipomœa. Ephemeral flowers. Supervolute æstivation. The disk. Use of pollen; —nectar. Septifragal dehiscence. Albumen. Vitality of seed. The bud. CONVOLVULACEÆ	182
LI.	ROCK MAPLE—Acer. Tree picturesque. Theory of leaf-forms. Autumnal colors. Maple sugar. Other Maples.	188
LII.	THE HORSE CHESTNUT—Æsculus. History. Phyllotaxy. Digitate leaves. Suppression of ovules. The SAPINDACEÆ. The Soapberry tree	192
LIII.	THE SILK GRASS—Asclepias. Pollinia. Corona, hoods and horns. Cross-fertilization. The ASCLEPIADACEÆ. The Cow tree. Carrion flower	195

CHAP.		PAGE
LIV.	SPOTTED KNOTWOOD—Polygonum. Ochreæ. Apetalous flowers. An achenium. Other species. The nectar defended from ants The POLYGONACEÆ	199
LV.	THE SPURGES—Euphorbia. A monandrous flower. A glandular involucre. Poisonous juice. The EUPHORBIACEÆ. Tapioca. Caoutchouc	204
LVI.	THE WHITE OAK—Quercus. Aments. Wind-fertilization. Acorns. Germination. History. Straightveined leaves. The CUPULIFERÆ. The value of Mast. Oak timber. Nut-galls	207
LVII.	THE WHITE PINE—Pinus. Acerous leaves. Triple pollen grains. Biennial fruit. The cone. Naked seeds. Root fences. Other species	214
LVIII.	THE HEMLOCK—Abies. Excurrent trunk. The CONIFERÆ. Pitted cells. The Douglass Fir. The Giant Cedars. Turpentine	218
LIX.	THE PALMETTO—Sabal. Tree with one bud. Caudex. The endogenous structure. Other Palms. Germination of the Cocoanut. The PALMACEÆ. Date Palm. Sago. Vegetable Ivory. Palm oil	223
LX.	JACK-IN-THE-PULPIT—Arisæma. The spadix and spathe. Golden Club. Calla. The ARACEÆ. Sweet Flag	229
LXI.	THE SHOWY ORCHIS—An orchidaceous perianth. Gynandrous stamens. Pollinia. How fertilized. A walking plant	233
LXII.	ORCHIS PSYCHODES. O. orbiculata. Lady's Slipper. The ORCHIDACEÆ. Mimicry of insects, birds, etc	237
LXIII.	IRIS, or BLUE FLAG. Ensiform leaves. Pollenization. The Fleur-de-lis in history	241
LXIV.	BLUE-EYED GRASS—Sisyrinchium. The IRIDACEÆ	244
LXV.	THE TRILLIUMS. Net-veined endogens. Wake-robin. The Bath-flower. Indian Cucumber. The TRILLIACEÆ	246
LXVI.	BELLWORT, or WILD OATS—Uvularia. Perfoliate leaves. Loculicidal pods. U. sessilifolia, and other species	254
LXVII.	LILY-OF-THE-VALLEY—Convallaria. Gamopetalous. Origin of the stem. History. Clintonia. The LILIACEÆ.	254
LXVIII.	THE STAR GRASS—Hypoxis. Sagittate anthers. Species.	258
LXIX.	THE SEDGES. GALINGALE—Cyperus. The umbels. The naked flowers. The spikes. The Rushes. Achenium.	263
LXX.	THE SEDGES. CAREX. Triangular culm. Monœcious spikes. Perigynium. Glume. The CYPERACEÆ. Papyrus	265
LXXI.	THE GRASSES. The turf. Culm. Sheath. Ligule. Poa pratensis. Wind-fertilization. Blue Grass. Fowl Meadow	271
LXXII.	ORCHARD GRASS—Dactylis. Secund panicles. Keeled glumes	274
LXXIII.	SWEET VERNAL GRASS—Anthoxanthum. Germination of the Grasses. The order GRAMINEÆ. The cereals. Bamboo. Hay—of what grasses made	276

> "*Happy, in my judgment,*
> *The wandering herbalist, who clear alike*
> *From vain, and that worse evil, vexing thoughts,*
> * * * * *peeps round*
> *For some rare flow'ret of the hills, or plant of craggy*
>   *fountains."*
>
> WORDSWORTH

> "*Flower in the crannied wall,*
> *I pluck you out of the crannies;*
> *Hold you here, root and all, in my hand,*
> *Little flower, but if I could understand*
> *What you are, root and all, and all in all,*
> *I should know what God and man is."*
>
> TENNYSON.

BOTANY.

I. PIGEON-WHEAT MOSS.

Description.—The portrait is before us (p. 15). We cannot fail to recognize a little rustic friend we have often met in our country rambles, covering the dry knolls in pastures among rocks and stumps.* Examining this plant as a whole, we see that it is a mossy herb, erect, 3 to 8 inches high, branched at the base, above which it is not branched, but *simple,* as the botanists say. Coming next to study the plant in detail, we find that it is *organized,* i. e., made of coöperating parts. Five of these—the root, stem, leaf, stalk, and fruit—are readily distinguished.

Analysis.†—The *Root* is the base of the plant. It grows in the ground, and is the part first formed. Its use is to hold the plant in its place and to take up nourishment from the soil. There are numerous small fibers or *rootlets* branching from the main root or axis, to give a broader foothold and aid in absorbing food.‡

The *Stem* (*a b*) springs from the root. It is upright, simple, 1 to 3 inches high, round and tapering, or, as we may hereafter say, *terete.*

---

* Specimens of this plant may be collected at all seasons of the year, and in drying they will lose none of their comeliness.

† *Analysis.*—Greek *ana,* each, or severally, and *luein,* to loosen or dissolve, means to consider anything in its different parts separately, one by one.

‡ The life-history of the Moss begins in a mesh of green, gossamer-like threads that spring from the spore which serves for its seed. For a time, 5 to 20 days, this mat slowly gathers strength, when suddenly on one of the crossings a tiny bud appears—a whorl of scaly leaves. Now true roots creep down into the soil, the threads drop away, the stem ascends. No one plants the Moss; it does not follow the track of man in his migrations; yet it is everywhere present to greet his coming. In the barren sands, in the chinks of the naked rock, wall, or pavement, wherever a spore may find a lodgment, there the Moss weaves its tiny mesh, and grows its diminutive forest.

The *Leaves* are green or brownish, and grow mostly from the upper part of the stem. They are (1) long-pointed, and shaped somewhat like a spear or lance, only narrower, approaching the form of a line ———; hence they are said to be *linear-lanceolate*. The edges are *serrulate* (Lat. *serrula*, a little saw).

The *Stalk* (*b c*) rises from the top of the stem among the leaves, and is therefore said to be *terminal;* and as it supports the fruit, it is called the *pedicel* (*pediculus*, a little foot). It is erect, usually longer than the stem, smooth, brown, thread-shaped (*filiform*).

The *Fruit* (*c*) is borne aloft on the pedicel. It is the last part produced by the plant, the chief end and aim of its whole life. It is a small square box or *capsule* (7), covered while growing by a hairy cap or *calyptra* (6). When ripe, the calyptra vanishes, the capsule nods (8), and the lid—*operculum* (9) at the top opens, revealing within a greenish dust. This consists of a multitude of tiny grains or *spores*, soon to be scattered on the ground, and to serve as seeds. After the operculum has fallen off, the mouth (*stoma*) remains open, and is seen beautifully bordered by a circlet of teeth. This is called the *peristome* (Gr. *peri*, around, *stoma*, mouth). With a microscope we can see that the teeth are blunt (10), and 64 in number.*

Classification.—At the top of the stem, before the fruit arises, we may often find a group of organs resembling and indeed serving as a flower (2).† But being colorless and

---

* In other kinds of Moss the number of the teeth is 4, 8, 16, 32—always some power of 2. Sometimes the peristome is double, the inner one consisting of as many little hairs (*ciliæ*.)

† In Fig. 1 (2) the artist has delineated a male, or sterile flower of Polytrichum (for two kinds of flowers are developed by this Moss). In 4 is seen, greatly magnified, the special organs (two *antheridia*, and *o, o*, two *paraphyses*) of the sterile flower. In 5, also greatly magnified, are seen the two *pistillidia* of a fertile flower, from one of which the capsule arises, the other proving abortive; 3 is one of the leaves (sepals).

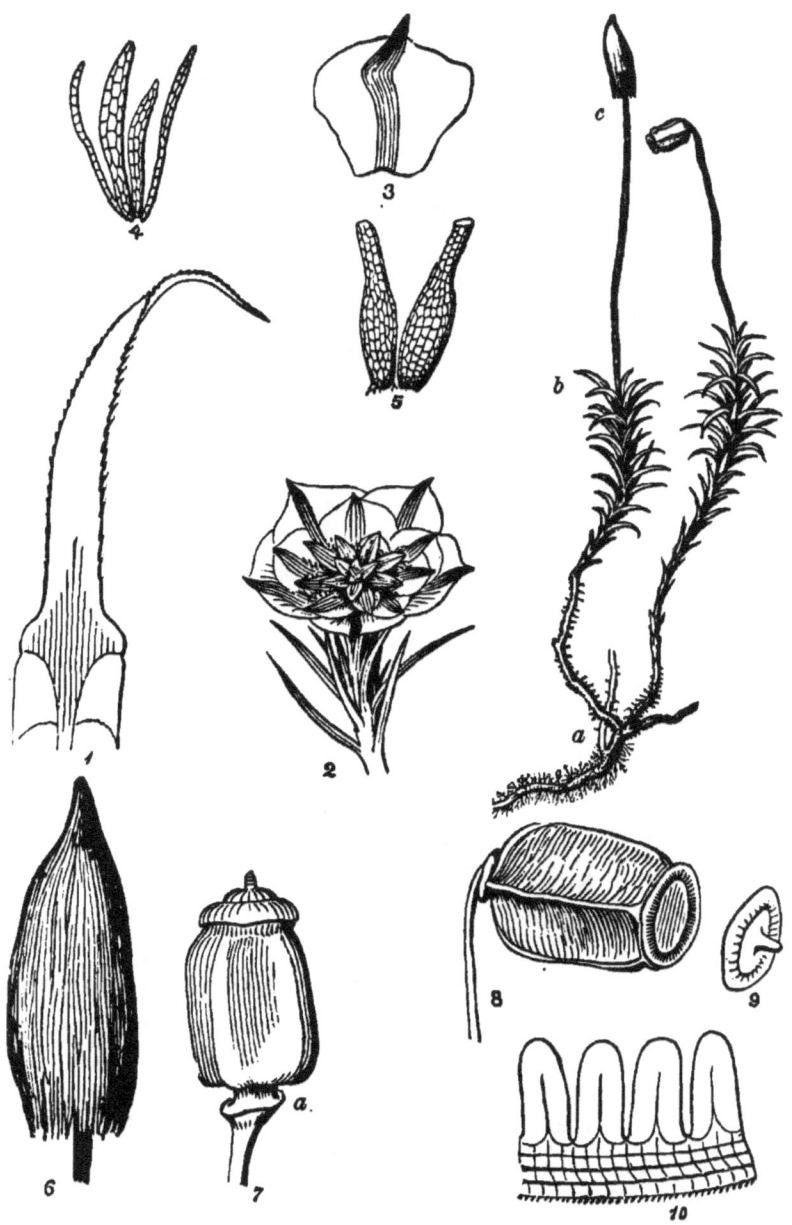

Fig. I.—Pigeon-Wheat Moss, Polýtrichum commùne. The portrait represents the plant in nearly full size. The other figures are *dissections*, showing the various organs magnified, as under a microscope; 2, a sterile flower, magnified.

insignificant in appearance, it is easily overlooked. Hence the early botanists called this and all the Mosses, Lichens, &c., CRYPTOGAMS (= Cryptogamia), that is, plants with hidden flowers, or more familiarly, the FLOWERLESS PLANTS.

The Name given to this plant by Linnæus,* the founder of the science of Botany, is *Polýtrichum,* a name derived from the Greek *polys,* much, *thrix,* hair; on account of its hairy calyptra.

ANALYSIS OF THE PIGEON-WHEAT MOSS.

Parts, Members, Organs.	DESCRIPTION.
THE PLANT.	*An herb 5 to 8 inches high, with leafy verdure, branched at base.*
ROOT.	*The basis of the plant, growing downward in the soil.*
AXIS.	*Crooked, diminishing downward.*
FIBERS.	*Many, short, branching from the axis.*
STEM.	*One or more from the root, erect, terete, 1' to 3', leafy.*
PEDICEL.	*Terminal, smooth, brown, thread-shaped, long, naked.*
LEAVES.	*Greenish, linear-lanceolate, pointed, crowded above.*
FRUIT.	*Terminal, erect at first, finally nodding.*
CALYPTRA.	*A cap of matted hairs, pointed at top.*
CAPSULE.	*A small box, generally four-sided, opening at top.*
Operculum.	*The lid of the capsule, round, pointed in centre.*
Peristome.	*The fringe of the mouth, consisting of 64 teeth.*
SPORES.	*Greenish, dust-like, contained in the capsule, for seeds.*

CLASSIFICATION. The Grand Division, **CRYPTOGAMIA.**
The Tribe, MOSSES or MUSCI.
Family or Genus, **Polytrichum.**

---

* In his botanical tours in Lapland, Linnæus found this Moss very abundant, and tells us that in his hours of rest he often made it his couch and pillow,

The Record.—In the preceding tablet are recorded the principal facts we have now learned concerning the Pigeon-wheat Moss.

Review of the Scientific Terms which have been employed and defined in this lesson. If the student will master them here, they need not be explained hereafter. Analysis. Axis. Calyptra. Capsule. Cryptogamia. Lanceolate. Linear. Operculum. Organized. Pedicel. Peristome. Rootlet. Simple. Spores. Stoma. Terminal. Terete.

II. THE APPLE MOSS.

Description.—This pretty Moss is known by its round, apple-shaped capsules. It grows in large, dense tufts, 2' or 3' (inches) high, of a light or yellowish-green color, often covering the ground on shady banks or in open woodlands.

Analysis.—The *Root* is a simple axis, clothed with minute rootlets, which appear like a soft brownish down.

The *Stems* are densely crowded, repeatedly forking, or *dichotomous* (dividing by pairs), covered and concealed by their leaves.

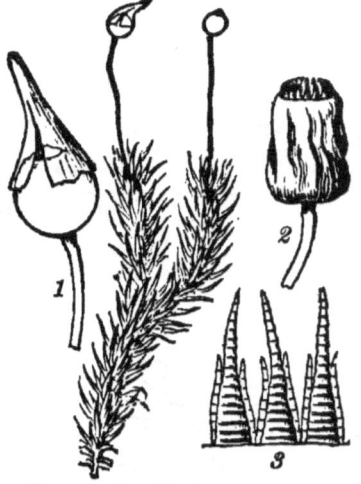

Fig. II.—Bartràmia pomifórmis.

The *Leaves* are numerous and crowded on the stem and branches. They are narrower than those of the Polytrichum, narrower even (proportionately) than a cobbler's awl; hence we define them as *linear-subulate* (*subula*, an awl).*

---

* Viewed under a strong microscope, the leaf of this Moss, and indeed every other part of it, appears a tissue of cells all of one shape and size throughout—polygons somewhat longer than wide. In other words, the Moss is wholly composed of *cellular tissue*.

The ***Stalk*** or pedicel is terminal or nearly so, erect, 6″ to 10″ (seconds = lines or twelfths of an inch) high, slender, yellowish, much shorter than the stem.

The ***Fruit*** or capsule (1) is slightly nodding, globular when fresh, oval and showing many ribs or furrows when dry (2). The cap or calyptra (1) is small, smooth, split on one side, and soon vanishing, or *fugacious* (*fugere*, to flee away). The lid or operculum is very small, somewhat conical. Under a microscope the peristome shows an outer row of 16 teeth, three of them being seen in the figure (3). There is also an inner row of as many hairs (*ciliæ*).

The Name by which this Moss is known in science is *Bartràmia*. It was conferred by Linnæus, A. D. 1750, in honor of John Bartram,* of Philadelphia. But this, however, is the title of a family or genus, including several kinds or *species*. A second name is therefore added to designate the species,† viz., Bartramia *pomifórmis* (Latin. *pomum*, apple; *forma*, form).‡

The Record.—Following the example given in the preceding lesson, the student will now fill the blanks in the annexed tablet. The descriptions are to be drawn from the text or from fresh observations of the plant (p. 17). See directions in "Suggestions to Teachers," p. 6.

The Order.—The Mosses are among the higher orders of the cryptogams. They have proper stems and green

---

* *Bartram* was a Pennsylvania farmer, said by Linnæus to be the greatest natural botanist then living. He traveled through the forests which at that early day covered so large a part of our country, collected plants and established in Philadelphia the first Botanic Garden in America.

† The same is true of our Polytrichum, whose specific name is *Polytrichum commùnè* (= common), or *P. commùnè*.

‡ Many object to scientific names in an elementary book. It should be remembered, however, that they are brief, exact, and universal; i. e., they are used in all scientific books and are known to all nations. The common names are local, and vary not only in different countries, but in different parts of the same country. In this work the English name is given first, then the classical or scientific. The student should know both, but in conversation may use either.

ANALYSIS OF THE APPLE MOSS.

Parts, Members, Organs.	DESCRIPTION.
THE PLANT.	
ROOT.	
Axis.	
Fibers.	
STEM.	
Stalk.	
LEAVES.	
FRUIT.	
Calyptra.	
Capsule.	
Operculum.	
Peristome.	
Spores.	

CLASSIFICATION.—Grand Division, or Subkingdom,..........
Tribe, or Order,..........
Name.—Genus and Species,..........

leaves, while the Moulds, Mushrooms and Lichens have neither. Not less than 2000 species have been described, chiefly inhabiting the cool and rocky regions of the Earth. On the cinders of Mt. Hood they form the first verdant specks of vegetation, and the cliffs of Mt. Washington are already green with mossy tufts and beds. Cold swamps are everywhere being filled with Sphagnum and other Mosses, whose remains accumulate and are, in time, condensed to peat—a valuable fuel in some countries where wood and coal are scarce.

Review of the Scientific Terms used in this lesson : Cilia. Dichotomous. Fugacious. Subulate. Species.

III. THE POLYPOD FERN.

Description.—This comely Fern is found everywhere in old forests, growing on stony steeps, and covering the rocks and boulders with a matted turf composed of their tangled stems and roots.

Analysis.—We may conveniently divide this plant into two portions or *regions;* first, that of the stem and root under ground (subterranean); second, that of the leaf and fruit above ground (aërial). The analysis will then proceed as before.

The *Root* consists of a number of thread-like or hair-like fibers, branching into tiny *fibrils*, growing all along the stem.

The *Stem* creeps about in the soil. It is many-branched, and covered with soft, narrow scales. As it never rises into the air it is properly a root-stock or *rhizome.**

---

* Plants with rhizomes are not uncommon. They are always *perennial*, i. e., living from year to year, and otherwise noted for their strong vitality. Those which have rhizomes long, slender, branching, are inclined to take exclusive possession of the soil, and so become in fields and gardens troublesome weeds. Such is the Polypod in Europe. On the contrary, in sandy sea-shores and dykes they are useful, binding the soil into a firm turf resisting the washing of the waves. *See* XIII and LXV.

POLYPODIUM.

The *Leaf* is all of the plant that is seen while growing. It is more than a mere leaf, since it bears the fruit as well as the foliage. Being thus a combination of stalk and leaf, it is called the *frond* (Lat. *frons*, a leafy branch). It stands inclined so as to present an upper and an under surface, both being green. At the base its stalk is called the *stipe*. Its margins are deeply cleft in numerous segments termed *pinnæ* (wings). Hence the frond is said to be *pinnatifid* or wing-cleft.

The *Veins* or ribs demand a careful study. There are three kinds. The *midvein* is the largest; it is the continuation of the stipe from the *base* of the frond to

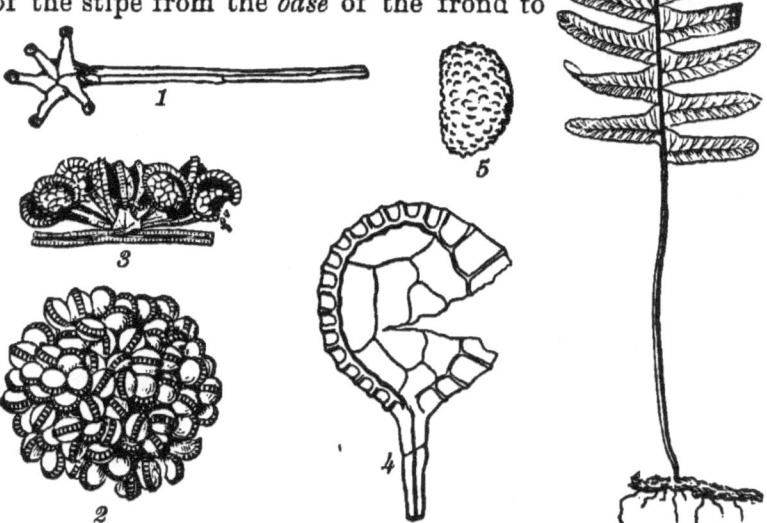

Fig. III.—Polypòdium vulgàre: 1, a fruit-bearing vein; 2, fruit-cluster; 3, a side view; 4, a capsule open; 5, a spore, greatly magnified.

its end or *apex:* 2d, the *veinlets* branch from the midvein and pass through the middle of each of the pinnæ: 3d, the *veinulets* branch from the veinlets, then divide or fork, and so end either in the edge (margin) of the frond, or in a fruit-cluster (2). This kind of veining (*venation*) in the larger veins is styled *pinni-veined*, or *feather-veined*, and that in the veinulets—*fork-veined*.

The *Fruit* is formed at the end of veinulets (1), on the under surface (the back) of the frond.* It consists of numerous round, reddish brown, regularly arranged patches called *sori* (*sorus*, plur. *sori*, a heap). Under the microscope these heaps (2) are found to be composed of numerous roundish vessels (*sporangia*), each on a pedicel (3) and filled with spores. The contrivance for opening these spore cases is very curious. Each one is clasped by an elastic, vertical ring. When ripe and dry, the ring contracts, breaks asunder, tears open the cell (4), and throws the spores (5) to a distance.†

Classification.—No flower is ever seen on this or any other Fern.‡ Therefore they are classed with the Mosses in the subkingdom CRYPTOGAMIA, or FLOWERLESS PLANTS.

The Name.—Polypod or Polypody is a contraction of

---

* The spores of the brake are hidden under the margin of the leaves, so that anciently it was thought that the Fern bears no seed. Later it was believed that the fern-seed was visible only on St. John's Eve, just at the moment when the saint was born :

"But on St. John's mysterious night,
Sacred to many a wizard spell,
The hour when first to human sight
Confest, the mystic fern-seed fell."

The superstitious belief that he who could at that hour get some of the fern-seed, became invisible, is frequently alluded to by the old poets. Shakespeare says :

"We have the receipt of fern-seed ; we walk invisible. '

† The spores of the Ferns are numerous. Let the student calculate them in one of these fronds. Professor Lindley observes of the Hart's-tongue (Scolopendrium), a small Fern, that each frond produces about 80 fruit-clusters (sori), with an average of about 4500 spore-cases in each cluster, and in each spore case 50 spores. The number of spores on each frond would then be $80 \times 4500 \times 50 = 18,000,000$ If all should grow, they would in a few years cover the whole continent.

‡ In germination, the spore of the Fern first develops into a green body resembling a Liverwort, called the *prothallus*. On this prothallus are certain little organs analogous to stamens and pistils, by which a *second set of spores* is generated, in advance of the true Fern. Thus in the Fern, as in some insects, there is an *alternate generation ;* it is first a Liverwort, then a Fern. (See *Zoology*, p. 220, Aphidæ.)

the Latin name, *Polypòdium* (Gr. *polys*, many, *poda*, feet), given it by Linnæus in allusion to its numerous creeping underground stems. As there are many species, ours is named *P. vulgàre*, the Common Polypod.

Scientific Terms defined in this lesson: Aerial. Fibrils. Fork-veined. Frond. Midvein. Pinna (plur. pinnæ). Pinnatifid. Pinni-veined. Prothallus. Rhizome. Root-stock. Sorus (plur. sori). Sporangium (plur. sporangia). Stipe. Subterranean. Veinlet. Veinulet. Venation.

ANALYSIS OF THE POLYPOD FERN.

Regions, Parts, Members, Organs.	DESCRIPTION.
SUBTERRANEAN.	*Parts under ground.*
Root.	*Many thread-like fibers and hair-like fibrils.*
Stem.	*Creeping, branching, scaly.*
AERIAL.	*Parts above ground.*
Frond.	*Pinnatifid, with oblong pinnæ, 8' high.*
Stipe.	*Green, naked (not scaly).*
Veins.	*The frame-work.*
Midvein.	*Largest vein, from base to apex.*
Veinlets.	*One in the midst of each lobe, or pinna.*
Veinulets.	*Forked, bearing the fruit.*
Fruit.	*On the back of the frond.*
Sori.	*Naked, rounded.*
Sporangia.	*Roundish, stalked, with an elastic ring.*
Spores.	*Yellow, dust-like particles, as seeds.*

CLASSIFICATION.—Subkingdom, CRYPTOGAMIA.
Tribe, The FERNS.
NAME.—Genus, *Polypòdium*; Species, *P. vulgàre*.

IV. THE OSMUND FERNS.

That tall Fern
So stately, of the queen Osmunda named:
Plant lovelier, in its own retired abode
On Grasmere's beach, than Naiad by the side
Of Grecian brook. WORDSWORTH.

Description.—Excursions in the old mossy damp woods, in the month of May, are generally rewarded by a rich display of these large, majestic Ferns already in fruit. They grow in circular clumps springing from a thick subterranean stem, or root-stock. As in the Polypods, the fronds constitute the aerial region. They are very smooth, often 3–5 feet in height, and a yard in width.

Clayton's Osmunda, shown in the cuts, bearing its fruit in the middle of the frond, is earliest ripe, and therefore the first to be analyzed.

Analysis.—The Root is of many fibers, with branches innumerable, short, spreading at right angles (*divaricate*), filling the soil.

The *Stem* is wholly subterranean, a thick blackish rhizome of loose texture and partly woody, living many years (*perennial*).

The *Frond* is twice divided; first into many distinct pinnæ, arranged in pairs along the lengthened stipe or *rachis* (Gr. back-bone); then each pinna is cut into oblong lobes or segments. This twofold division is termed pinnate-pinnatifid or *bipinnatifid*.

The venation, like that of Polypod, is pinni-veined and fork-veined (2).

Vernation.—When starting from the ground in early spring, each frond is a coil rolled from the top inward and downward, gradually unfolding, scroll-like, as it grows (5).

FIG. IV. Osmŭnda Claytoniàna, with five dissections; 2, a segment or lobe, showing the forked venation; 3, clusters of spore-cases (enlarged); 4, one of the spore-cases (capsules) still further magnified; 5, young frond, showing its mode of vernation.

NOTE.—Specific names are generally *adjectives*, and should never begin with a capital letter, except (1) when the name is derived from a person or a country, as O. Claytoniàna, or Erythrònium Americànum; and (2) when it is a *noun*, as Dodecàtheon Meàdia.

This mode of bud-folding* is termed *circinate* (Lat. *circinus*, a compass).

The *Fruit* is densely clustered on some of the middle pinnæ, which contract their leaf-portion to a mere green edge. The sporangia or spore-cases open lengthwise into two halves or valves, containing the dust-like spores.

The Name of this noble genus of Ferns is *Osmúnda*, from Osmunder, one of the titles of the Celtic Thor, god of thunder, whence we also derive the word Thursday (Thor's-day). The species just analyzed is *O. Claytoniána*, named in honor of John Clayton, one of our earliest botanists.

O. cinnamómea, the Cinnamon Fern, has some of the central fronds of each clump wholly fertile and condensed into fruit, tall cinnamon-colored clusters, looking like flames —whence they are often called the Flaming Fern.

O. regális, Royal Osmund, our tallest Fern, with its fronds separated into innumerable distinct leaflets, and its fruit all terminal, is celebrated for its regal beauty.†

The Order.—Ferns constitute one of the tribes or orders of the Cryptogams, named in science FILÌCES. They grow in all countries, but are most abundant in New Zealand and the tropical islands, where the climate is warm and damp. Of the 2000 species known, not more than 200 are native outside the tropics, and not over half of these in the United States.

The Ferns are the largest of the Cryptogams. In the Tropics they become trees, their pinnated plumes drooping

---

* The term *vernation* (*vernus*, the spring) was invented by Linnæus to express the general idea of bud-folding.

† In Europe this Osmund grows to a height often of 11 feet, and its great masses of green leaf-sprays form a marked feature in the landscape. Its tall stalk generally stands erect, but sometimes it acquires a drooping habit, as at the Lakes of Killarney. It there fringes the banks, especially of the river which connects the lakes, and its long fronds arching gracefully over, dip into the crystal water, forming coverts whence the birds gaze fearlessly out upon the passing traveler.

from the summit of trunks 40 feet in height. Fern remains are abundant in bituminous coal (see *Geology*, p. 155), and the rocky roofs of the mines are frescoed with the delicate tracery of their fronds in wonderful variety. These fossils indicate that at one period of the Earth's history, the Ferns constituted a large part of the vegetation.

Among our native Ferns are the Maidenhair (Adiántum), the Climbing Fern (Lygòdium), the Common Brake (Pteris). Splendid tropical Ferns flourish in our conservatories. Many of our own may be easily cultivated in the open air, or in Wardian cases with pleasing results.

The Ferns are not important either as food or medicine. The rhizomes and young shoots of several species are eaten in Australia and Oceanica in the absence of better food. Aspídium fragrans has the scent of raspberries and has been used for tea. From the Maidenhair a cough syrup called Syrup of Capillaire, is made. Aspídium Filix-mas is an effectual vermifuge.

The Cryptogams.—Besides the Ferns and the Mosses, this Grand Division includes also the Clubmosses (Lycopods), the Horsetails (Equisetaceæ), the Liverworts (Hepaticæ), the Scalemosses (Lichens), the Seaweeds (Algæ) and the Moulds and Mushrooms (Fungi), plants descending to the lowest rank, the simplest structure, and the minutest dimensions.*

Scientific Terms defined :—Bipinnatifid. Circinate. Divuricate. Perennial. Pinnate-pinnatifid. Rachis. Vernation.

---

* The Cryptogams are numerous and minute beyond conception. They inhabit every clime, from the Equator to the Poles. They lie at the foundation of all life. Without them vegetable and consequently animal life would be impossible. They —their lower tribes—are the first to grow on cinders, sands and rocks. The last they gradually disintegrate, and, by the decay of successive generations, form a' length a soil capable of sustaining plants of higher orders—grains, grasses and trees on which animals may live. Thus plants of higher rank replace those of lower, and fatten on their spoils. But sooner or later these also perish, and then the Cryptogams resume their sway. On fallen leaves and trunks they multiply, encompassing penetrating, consuming, and in a few years restore to the earth, with interest, the materials which they had borrowed.

ANALYSIS OF A FERN.

Regions, &c.	DESCRIPTION.
SUBTERRANEAN.	
Root.	
Stem.	
AERIAL.	
Frond.	
Stipe.	
Rachis.	
Veins.	
Midvein.	
Veinlets.	
Veinulets.	
Fruit.	
Clusters.	
Sporangia.	
Spores.	

CLASSIFICATION.—Subkingdom,...............
Order,...............
NAME.—Latin,............... English,...............

V. THE DOG-TOOTH VIOLET.

Description.—Spring has come again. The winds blow soft from the West and South over the melting snowbanks. Birds once more fill the air with song, while the plants, awakened from their winter's sleep, put on their robes of leaf and flower. Down in the woody vale, or in the thicket by the river, the Dogtooth Violet already hangs out its yellow bell. Though scentless, the flower attracts by its airy grace. We must dig carefully around its tender stalk if we would raise it entire, for its root strikes deep into the loamy soil. Examining the plant as a whole, we find it smooth and polished in surface, plump and fleshy in substance, and plain in outline. As it lives above-ground only one season, dying at the approach of Winter, it is an *herb*.

Analysis.—The whole plant may be divided into two parts—the *Leaf region* and the *Flower region*, and each of these again into three parts. To the Leaf-region belong the root, stem and leaf; to the Flower-region, the stalk, flower, and fruit. A little reflection will show that the former parts work for the plant itself, and the latter for its posterity which is to spring from its seed.*

THE LEAF REGION.—The *Root* is fibrous, i. e., it consists of *fibers* and *fibrils*. The former start from the bottom of the stem deep in the ground, and are long and white; the latter are the minute subdivisions of the fibers.

The *Stem* is a simple, slender column (*caulis*) with its lower end apparently enlarged into a bulb, whence it is called a *bulbous stem*. The *bulb*, which is egg-shaped or *ovoid* (Lat. *ovum*, an egg), consists of many scales, thick, white, and

---

* Hence the former are called the *vegetative* organs, and the latter, the *reproductive*.

THE DOG-TOOTH VIOLET.

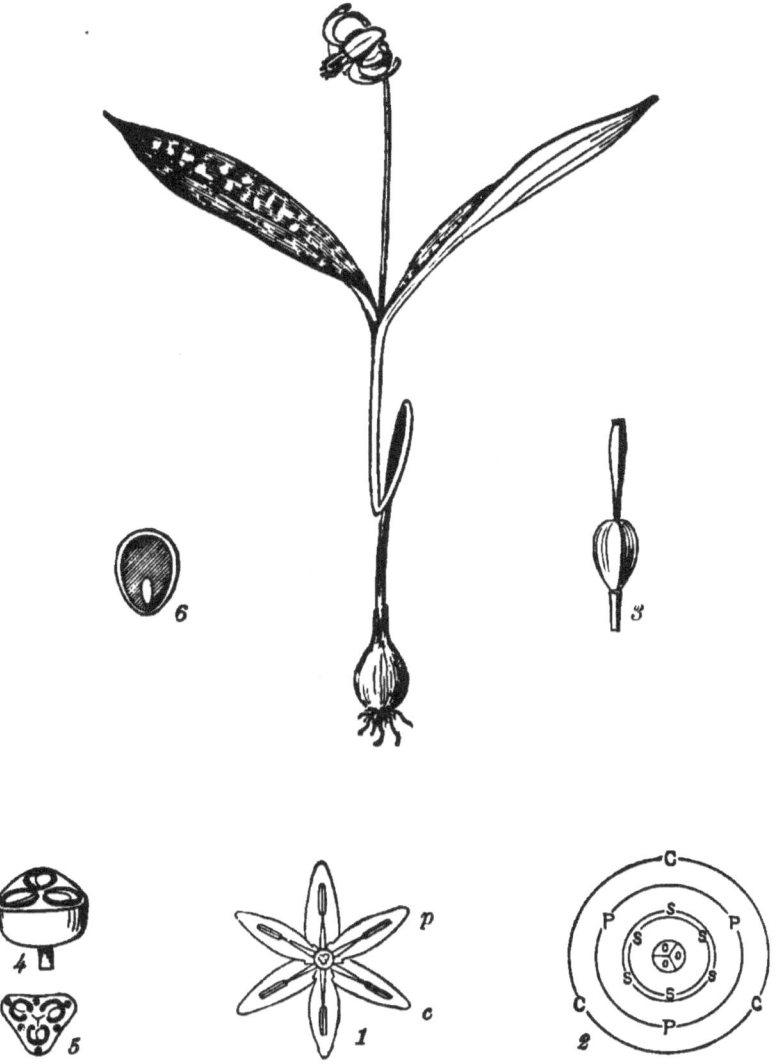

Fig. V.—Erythrònium Americànum: 1, an expanded flower; 2, plan of the flower; 3, the pistil; 4, 5, sections of the fruit; 6, section of a seed, showing the embryo and albumen.

broad, growing out of the solid base from which the stem arises.

The *Leaves*, 2 in all, are placed nearly opposite each other on the stem, the lower being the larger. Their rich green color is singularly variegated with purplish and whitish blotches. They stand out on a narrow base—a footstalk. This is called the *petiole*, and the expansion of the leaf is the blade. The latter is in outline both elliptical and lanceolate, so that we may style it *elliptical-lanceolate*. The apex is *acute*. The margins are even and *entire* (without teeth or notches).

Venation.—The leaf is marked with fine lines running lengthwise in nearly parallel curves corresponding with the contour of the margin. They indicate the course of the *veins* and show what is called a parallel venation.

THE FLOWER REGION.—The *Stalk* which supports the flower is called the *peduncle* (a term higher in rank than *pedicel*). Its top, forming the basis of the flower, is the *torus*. It bends under the weight of the flower, but bears the fruit erect.

The *Flower* is the latest, gayest, and frailest part of the plant. It is *solitary*, drooping or *nodding* from the curved top of the peduncle. We can easily count 13 distinct members or organs composing it. There are 6 lanceolate *recurved* (bent back) leaflets, and six slender columns, all standing on the torus around one central column. The 6 leaflets together constitute the *perianth* (Gr. *peri*, around, *anthos*, flower).*

These 13 may be assorted into 4 groups of organs. The 3 outer leaflets, (*c*) orange-red on the back, are *sepals*, and together form the *calyx* (cup, or goblet). The 3 inner are

---

* The leaflets of the perianth are recurved more or less according to the hour of the day and the amount of sunlight. At night they close, and gradually open as the day advances, and at midday in a bright sun they are recurved as shown in our cut. See XXI, on the Sleep of plants.

petals (*p*) and constitute the *corolla* (crown). Let the student take note of the two blunt teeth opposite each other near the base of each petal. Within the corolla and perianth are the 6 *stamens*—those slender bodies arranged in 2 circles. Each stamen is made up of a *filament* (*fila*, thread) bearing at the top an *anther*—a little oblong box with 2 cells opening lengthwise by slits. These contain minute grains of *pollen*, which in due time escape by the open slits.

The *Pollen* must be studied under the microscope. Then it appears composed of innumerable grains, oval, yellowish-brown, and peculiarly marked (7). Their use in aiding the formation of the seed will be shown in future lessons.*

The *Pistil* (*pistillum*, pestle) is the central organ (3). Being the most important part of all, destined to become the fruit and seed, it is thus surrounded and protected by all the other organs. It is composed of the *ovary* at the base, the stigma at the top, and the *style* between. The stigma is 3-lobed, and on dissecting the ovary (4, 5) we shall find 3-cells. May we not infer that there are 3 pistils united into this one compound pistil?

PLAN OF THE FLOWER.—The diagram (2) exhibits the relative position of the different members of this flower as they stand upon the torus. They occupy five circles. The outer circle is the calyx, and *v c c* the position of the sepals. The 2d circle is the corolla, and *p p p* the position of the petals. The 3d and 4th circles are the stamens, and *s* their

---

* Infinitely more pollen is produced than is needed, just as in spring time there are more blossoms on the apple tree than could possibly be matured. So abundant is the yellow pollen developed and shed by the Pine that we frequently see the ground in Pine forests covered with it; and swept off by the winds, it falls at a distance, alarming the ignorant with a "sulphur rain." The amount shed by the grasses fills the air, producing in sensitive nostrils the effect called the "hay fever." A single Pæony-flower, according to Darwin, produced 3,654,000 grains of pollen. Everywhere in the floral world we meet with the two opposite rules of profuse prodigality and extreme economy. "It is the moral of the New Testament story—feeding the hungry thousands and then gathering up the fragments that nothing be lost."

places severally. The inner circle shows the position of the 3 united ovaries (*o o o,* called also *carpels*). Each member alternates in position with its next neighbors; e. g., the petals with the sepals and the outer stamens, &c. And the members are all in 3s—3 sepals, 3 petals, 3 outer stamens, &c. Hence the flower is said to be *alternating, symmetrical, and 3-parted.*

The *Fruit* is the full-grown and complete ovary. The flower is of short duration. The sepals, petals, stamens and style soon fade, wither, and fall. They are *deciduous.* But the ovary is *persistent,* growing, and ripening a month after the flower has done its work. It is then a dry seed-box—a *capsule,* having its 3 cells packed full of seeds.

Here we come to the ultimate product of the plant, that for which it lived, grew, bloomed, and labored. In the shell of the seed (called the *testa*) is safely inclosed the rudiment of a new plant ready to be developed into the likeness of the parent. A careful dissection (6) will show the structure of this rudiment—the *embryo.* It appears a *simple* cylindrical body lying in a white fleshy substance called *albumen.* (See p. 42.)

Classification.— The Dogtooth Violet, by the presence of its conspicuous flower, is in marked contrast with the Pigeon-wheat and the Polypod. They represent the Flowerless (Cryptogamia) and this the Flowering Plants (Phenogamia).* Moreover, by the presence of *parallel-veined leaves, 3-parted flowers, and seeds with a simple (not lobed) embryo,* this plant represents the ENDOGENS, one of the two provinces into which the Flowering Plants are divided. (For the other province, see p. 43.)

The Name.—This plant is a Lily rather than a Violet,

---

* Thus the Vegetable Kingdom is parted into two subkingdoms, known by the absence or presence of visible flowers. This division was first recognized A. D. 1682, by John Ray, of England.

THE DOG-TOOTH VIOLET.

ANALYSIS OF THE DOG-TOOTH VIOLET.

ORGANS, &c.			DESCRIPTION.
THE PLANT.			A smooth, fleshy herb.
FLOWER REGION.	INFLORESCENCE		Solitary, terminal, pedunculate.
	FLOWER.		Drooping, bell-form, 3-parted, alternating.
	CALYX.		Orange, greenish, and yellow.
		Sepals.	3, lanceolate, recurved.
	COROLLA.		Within the calyx, yellow.
		Petals.	3, lanceolate, yellow, recurved, 2-toothed.
	STAMENS.		6, in two rows or sets.
		Filament.	Linear, yellowish.
		Anther.	Linear-oblong, 2-celled, yellow.
		Pollen.	Elliptical, yellowish, minute grains.
	PISTIL.		Triplex, and triply compound.
		Ovary	Green, 3-sided, 3-celled, persistent.
		Style.	Club-shaped or clavate, deciduous.
		Stigma.	At top of the style, 3-lobed.
	FRUIT.		An erect, compound capsule.
		PERICARP.	Dry, opening by 3-valves.
		SEEDS.	Many, ovoid, with a curved point.
LEAF REGION.	LEAVES.		2, nearly opposite, mottled.
		PETIOLE.	At the base of the leaf, linear.
		BLADE.	Elliptic-lanceolate, acute.
		VENATION.	Parallel.
	STEM.		Mostly subterranean, blanched.
		BULB.	At the base, ovoid, white, of scales.
		CAULIS.	Simple, 6-10' long, annual.
	ROOT.		Consisting of white fibers, deep down.

LOCALITY.—Damp shady woods, (Date) April 10th, 1877.
CLASSIFICATION.—Subkingdom, **PHENOGAMIA**.
   —Province, ENDOGENS.
   —Order, THE LILYWORTS.
NAME.—English, *Dog-tooth Violet.*
  —Latin, **Erythronium Americanum**.

as we shall see hereafter (p. 257); hence the common name is false as well as inelegant. The term Dogtooth may refer to the two indentures on each petal (p. 32), or to its bulbs as they grow in England. The scientific or classic name is *Erythrònium* (*eruthros,* red, the color it often assumes in Europe). The name of our species is *E. Americànum.*

Scientific Terms Defined.— Acute. Alternating. Anther. Bulbous stem. Calyx. Capsule. Carpels. Caulis. Corolla. Deciduous. Endogens. Entire. Fibrils. Fibrous. Filament. Herb. Ovary. Ovoid. Parallel-veined. Peduncle. Perianth. Persistent. Petals. Petiole. Phenogamia. Pistil. Pollen. Recurved. Sepals. Solitary. Stamens. Stigma. Style. Torus.

VI. THE TULIP.

Description.— The tulip is said to be a native of Mt. Taurus and the adjacent region. It has been cultivated by florists for more than three centuries in Europe, and one in America. It blossoms in the garden a week or two later than the wild Erythronium, with which it is closely related. Careful study will reveal between them striking differences as well as resemblances. Viewing the plant as a whole, mark its height, its *glaucous* (sea-green)* color, its smooth surface.

Analysis.—THE LEAF REGION.—The bulb, if cut across will exhibit a series of concentric rings, each one being an entire layer. Bulbs so constructed are called *tunicated,*† —a form familiar in the Onion. The Leaves are large

---

* According to the Greek Mythology, *Glaucus* was the name of a fisherman who leaped into the sea and "by transmutation strange" became a sea-god. Hence the botanists use the word to express the pale sea-green color of the foliage of certain plants, as well as of the whitish powder which sometimes covers them.

† The Lily grows from a *scaly bulb,* i. e., a bulb composed of scales, each forming but a partial (not entire) layer or ring. In the heart of the Tulip bulb, protected by these layers, are hidden not only the future stem, but the leaves also, and even the coming flower with all its various organs. This bulb also contains under its outer

and few, rarely more than 3. In margin, they agree with Erythronium except that they are wavy. In figure, they are partly lanceolate, and partly with the outline of an egg, being broadest just below the middle: hence they are *ovate*-lanceolate. They have a *clasping* base and no petiole. Compare the venation with that of Erythronium (p. 31).

THE FLOWER REGION.— Here note the *attitude* of the flower in contrast with that of Erythronium; but its organs are of the same number and kind.* The Sepals and petals are distinguished only by their position. Which are the outer? They are all broadly ovate, blunt (*obtuse*) at the apex, concave, and not recurved nor spreading. The ovary and its resulting pod is in the form of a triangular prism. The 3 stigmas

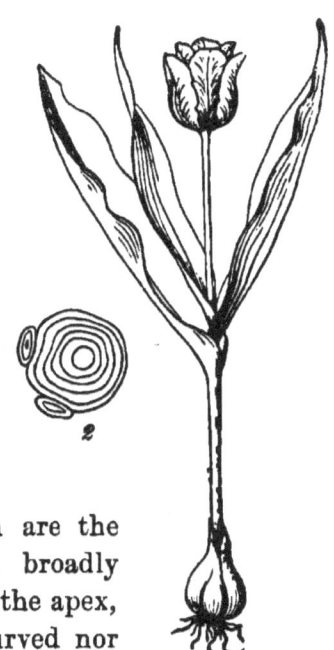

Fig. VI. Tùlipa Gesneriàna; 2, section of the bulb; 3, the pistil.

coat two minute buds ready to be developed in turn to take the place of the parent (2). Thus, after flowering and fruiting, we shall find the bulb which we planted shriveled and empty, having imparted its substance to its offspring, but already *replaced* by another or two, full grown from the buds, and replete with rudiments like the former. So provident is Nature. "No leaf drops till a new one is prepared to take its place; no flower perishes naturally till its house is made ready and furnished with seeds. In Autumn, the sad season of decay, there is yet as much of life as of death." Amidst the tokens of death are the elements of growth. In the autumn buds of the Oak are hidden its future leaves ; in those of the Lilac, its coming leaves and flowers. In the bulb of the Hyacinth, another season's blossoms are clearly seen even with the naked eye. The rich mucilage of the Slippery Elm, and the sweet sap of the Sugar Maple, are provided beforehand for feeding the young buds and hastening their early development. Thus within a few days a large tree will cover itself with foliage and bloom.

* The flower of the Tulip like that of the Rose (p. 114) is often *double*, in which state it is unfit for analysis.

are quite distinct, *sessile* (sitting) on the ovary, there being no style.

The Name is from a Persian word signifying a turban, whose gay colors it resembles. In Latin it becomes *Tùlipa*. The Common Tulip, here figured, is *T. Gesneriàna*, being dedicated to Gesner, a Swiss botanist, who saw it blooming in a garden in Augsburg and first made it public in 1559.*

Varieties.— This flower indulges in many freaks and fancies as to the cut and color of her robes. Now they are single, now double, and now semidouble. Here they are yellow, there white, and even crimson, purple or carmine. Again they are mixed—striped, spotted or flecked in endless combinations of colors. These are merely *varieties* of the same species, induced by their treatment in cultivation. Names are given them by the florists, rarely by botanists, such as Bizarres, Duc Van Thol, Bubloones, Comte de Pompadour, Parrotts, &c.†

The Record.—With these few hints to guide, let the student now complete the analysis of the Tulip and write its record in the annexed tablet.

Scientific Terms.— Clasping. Obtuse. Ovate. Scaly bulb. Sessile. Tunicated. Varieties.

---

* The taste for cultivating the Tulip spread into the Netherlands, and about 1634 increased to such an extent that all classes began to speculate in the bulbs. Houses and lands were sold to be invested in flowers. Ordinary business was neglected. Sudden fortunes were made. Nobles, mechanics and chimney-sweeps alike flocked to the tulip-market. Prices increased until a single bulb (the Semper Augustus) sold as high as $6,000 of our present money. A story is told of an English botanist who, traveling in Holland, happened to see a tulip-root in a conservatory. Ignorant of its value he began to peel off its coats to examine its peculiar structure. While immersed in his botanical study, the owner suddenly rushed in and in an agony of rage shouted "It's an Admiral Van der Eyck!" In vain the traveler protested his scientific intentions. He was dragged before a magistrate, where, to his consternation, he learned that the innocent-looking bulb was worth 4000 florins and that he was to be held in confinement until he found securities for that sum.—At last this tulip rage ran its course. Prices suddenly fell. The rich of yesterday became the poor of to-day. A commercial crisis ensued. Holland did not recover from the "Tulip mania" for many years.—The love for this flower still exists in that country. We import our best bulbs from Holland, and the wealthy Dutchman boasts of his fine tulips as a rich Englishman does of his horses or paintings.

† For the Order of Tulip and Erythronium, see LXVII.

ANALYSIS OF THE TULIP.

ORGANS, &c.	DESCRIPTION.
THE PLANT.	
Root.	
Stem.	
Leaves.	
Inflorescence.	
Flower.	
Calyx.	
Sepals.	
Corolla.	
Petals.	
Stamens.	
Anthers.	
Style.	
Stigma.	
Ovary.	
Fruit.	
Seed.	

Locality,............
CLASSIFICATION.—Subkingdom,............
—Province,............

Order............
NAME.—English,............
—Latin,............

VII. THE SPRING BEAUTY.

Description.—Early in Spring, in the grassy meadow, along the shady margins of the woods, or under tangled thickets, often in company with the Dog-tooth Violet, lo! the Spring Beauty! Its roots, like those of its neighbor, strike deep into the soil, and in order to lift the plant entire we must make careful use of the trowel. Viewed as a whole, the Spring Beauty in outline, surface, and substance, resembles the Dog-tooth Violet. It is also divisible into the same regions and parts.

Analysis.—THE LEAF REGION.—The *Root* is a new and singular structure. There is a massive body, irregularly rounded, brown without, white and starchy within. To this *tuber*,* as it is called, are attached the ordinary fibers and fibrils. These are the true working roots, absorbing plant-food from the ground, while the tuber serves as a reservoir for its reception after being digested and changed to starch.

The *Stem* is a simple, slender column, a few inches in height. In substance, it is *herbaceous*, that is, tender, juicy, flexible, greenish. In attitude, it is erect and upright; for, though weak, it stands unsupported.

The *Leaves*, 2 in number, grow opposite each other in the air and light at the top of the stem. In outline, they are narrowly lance-shaped, nearly as in the Pigeon-wheat, *linear-lanceolate*, or almost *linear*. They are fixed to the stem by their base without the intervention of a petiole; that is to say, they are *sessile* (sitting), while the upper end, the apex, as in all plants, is *free;* their margin is entire, and color, green.

THE FLOWER REGION.—The *Flower* is a curious gem,

---

* Botanists generally consider the tuber as a member of the stem. But to avoid subtle distinctions at this early stage, we here incline to the popular view regarding it as a member of the root.

40 THE SPRING BEAUTY.

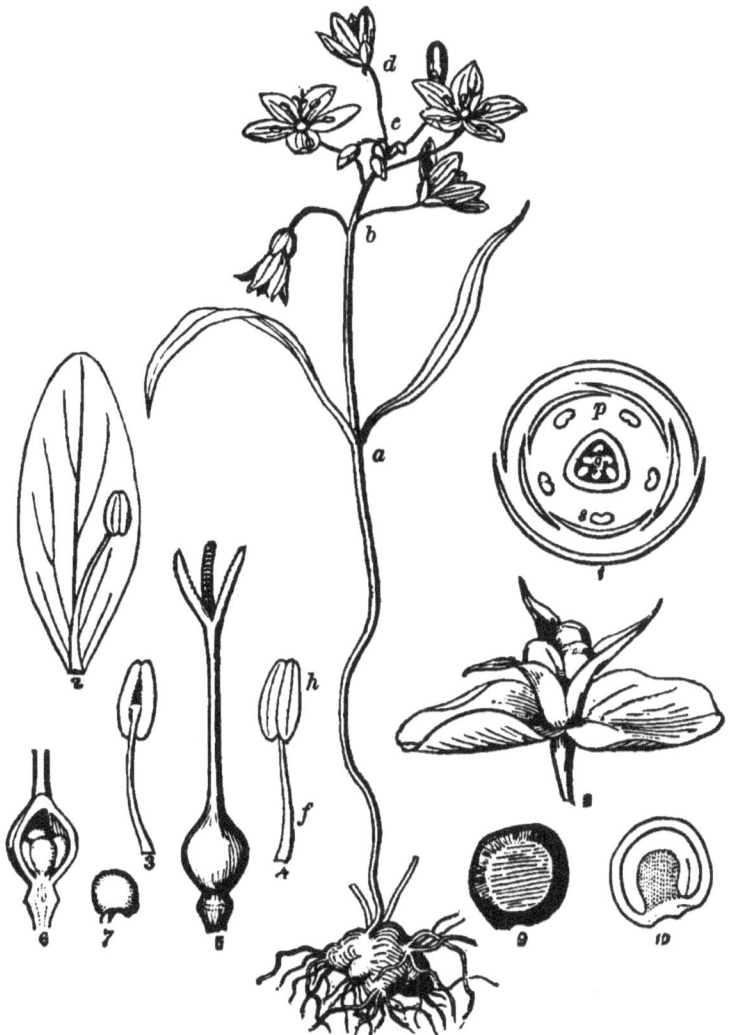

Fig. VII.—Claytònia Virgínica: 1, plan of the flower; 2, a petal with a stamen opposite; 3, 4, stamens, front and rear view; 5, the pistil; 6, ovary dissected; 7, an ovule; 8, fruit just opening, with calyx; 9, a seed; 10, the embryo.

inviting study, and losing none of its interest and loveliness by a close inspection. Let us first observe the situation and arrangement, or what the botanists call

The *Inflorescence*. The flowers form a cluster above the leaves at the termination of the stem. In the cluster appears (*a* to *b*) a general foot-stalk—the *peduncle*—supporting the whole; an axis (*b* to *c*), called the *rachis* (spine) running through the midst; and several special foot-stalks (*d*), *pedicels*, branching from the rachis, each bearing one flower. The whole cluster so arranged is a *raceme*.* The flowers do not all open at once, but in a regular succession, beginning with the lowest in the raceme and ending with the highest. While the lower are in bloom or past bloom, the upper are in bud or just opening. The word *centripetal* † is used to express this special mode of inflorescence.

The *Flower* is made up of four sets of organs, each set a circle one within another. First, the *calyx* or cup (*c*) containing all the rest. It consists of 2 green leaflets called *sepals*. Next within is the delicately colored *corolla* consisting of 5 rose-tinted and red-lined *petals* ‡ (*p*). Third, a circle of 5 stamens (*s*), each consisting of a slender filament (*f*) tipped with an oblong anther (*h*). We must not fail to observe their unusual situation—each opposite to (*opposing*) a petal (2). Observe also (3, 4) how the anther is attached to the filament, how its 2 cells open, and how it seems to face this way or that. Here it faces the pistil, and we say it is *introrse;* and the cells open *lengthwise*. And fourth, the pistil (5) in the center of the flower, consists of one ovary, one style and three stigmas (*g*). If we dissect the ovary (6),

---

* Sometimes in vigorous specimens the rachis divides, forming two racemes or a double raceme.

† The top of the inflorescence is regarded as its center, the base the circumference; hence the fitness of the word (*centrum*, the center, *peto*, I seek).

‡ The colored converging lines and veins which mark the petals of Spring Beauty, Pansy, Geranium and other flowers, serve as honey-guides for insects. They invariably converge towards the nectaries at the base of the petals and stamens. An insect following them is led directly to where the honey is secreted. On its way its body is dusted with pollen, or, already dusted, is brought into contact with a pistil ready to receive it.

we shall find within its cell 3 or 6 young seeds (*ovules*). Here also, as in the pistil of Erythrònium, are clear indications of tri-unity.

The *Fruit*. While the ovary is growing and ripening into fruit, it is attended and protected by the *persistent* calyx (8); but the corolla, stamens and style are early *deciduous*. The fruit is finally a dry seed-box or capsule (8), opening by 3 valves, and disclosing 3 or more black, shining, lens-shaped seeds (9).

The *Seed*. Here again we come to the ultimate product of the plant. The seed contains the curved embryo (10)—the young plant slumbering in its cradle. Surrounding it is a white mass of *albumen** (*albus*, white), a storehouse of food provided for the sustenance of the young plant after awaking and before its roots can draw nourishment from the soil. Mark here the structure of the embryo (in contrast with that of Erythronium), how it is curved, and cleft at the upper end into two equal seed-lobes, or *cotyledons*.†

Plan of the Flower.—The diagram (1) indicates the relative position of the organs as they stand on the torus; first, the 2 sepals; 2d, the 5 petals; 3d, the 5 stamens *opposing* the petals; 4th, the ovary—3 in 1. Why then is this flower *unsymmetrical?* Why is it 5-parted?

Æstivation.—This diagram also shows how the envelopes are folded in the bud, that is, their *æstivation* (*æstivus*, in summer); the buds themselves will show it much better. The margins do not exactly meet, but overlie each other like

---

* By a wise provision, the albumen is deposited in the form of starch, which is insoluble in water, or else the first rain might dissolve and waste the young plant's inheritance. There is, however, laid up also in the seed a bit of ferment called *gluten*. By the action of moisture this will slowly change the starch to sugar; and that being soluble can be used by the tiny shoot as it needs. (See *Chemistry*, pp. 184 and 194.)

† The ovules in the cut 6 are growing *erect* from the base of the cell; (7) shows an ovule more advanced, with its stalk (*funiculus*). Comparing this with (10), it is evident that the ovule in growing bent over on itself, bringing its apex near its base.

shingles on a roof, i. e., are *imbricated* (*imbrex*, a tile).* A special mode of imbrication is seen in the petals, of which 2 are wholly within, 2 wholly without, and 1 is partly both, having one edge within and one without. This is the *quincuncial* æstivation and very common.

The Name, *Claytònia*, was given by Linnæus to this plant in honor of *John Clayton*, who sent it to him, in 1757, from Virginia. Hence this species is called *C. Virgínica*. Another species was first seen by Michaux, about 1800, in the mountains of N. Carolina and so named *C. Caroliniàna*. But it is far more common northward from New England to Wisconsin. You may know it by its leaves being shorter and broader—elliptic-lanceolate. What other differences do you find?

The Order.—The Claytonias belong to the same order with the splendid flowering Portulacas, viz. PORTULACACEÆ, or the PURSLANES. That troublesome weed of the gardens —the common Purslane, is also a species of the genus *Portulàca*—*P. oleràcea*. Its small yellow flowers appear in Summer, and its curious seed-boxes in September. These open by a lid crosswise, and bear the classic name of *pyxis* (a box. Fig. XLIII, 5). In some countries Purslane is esteemed as a pot-herb, and a salad, on account of its cooling antiscorbutic properties (*Lindley*).

Classification. — In contrast with Erythronium, the genus Claytonia, and its order, by their *2-lobed embryo, and their flowers 5-parted* (*or at least not 3-parted*), represent the EXOGENS, the other province of the Flowering Plants (p. 33).

Scientific Terms.—Æstivation. Albumen. Centripetal inflorescence. Herbaceous. Imbricated. Introrse. Opposing stamens. Pedicel. Pyxis. Quincuncial. Raceme. Rachis. Sessile. Tuber.

---

* In other plants the sepals or petals may be found to meet edge to edge. Such æstivation is called *valvate*. Indeed the valves of the capsule of this plant thus meet while closed (8). See other modes of æstivation described in p. 85.

ANALYSIS OF SPRING BEAUTY.

The Record.—With the following tablet as a guide, let the student record the analysis of our other Claytonia, or of Purslane, or a Portulaca, in the annexed blank tablet.

Organs.	DESCRIPTION.
THE PLANT.	*An herb, terrestrial, 3—6' high, fleshy, smooth.*
ROOT.	*Brown fibers with a roundish, starchy tuber.*
STEM.	*Herbaceous, simple, upright.*
LEAVES.	*2, opposite, sessile, linear, or lance-linear, 3—6' long.*
INFLORESCENCE.	*Terminal raceme, with long pedicels, centripetal.*
FLOWER.	*5-parted, unsymmetrical, opposing, 8'' diameter.*
CALYX.	*2-parted, green.*
Sepals.	*Persistent, 2, erect, ovate, short.*
COROLLA.	*5-parted, delicately colored and pencilled.*
Petals.	*5, deciduous, spreading, obovate, roseate.*
STAMENS.	*5, opposing the petals.*
Filament.	*Slender, white.*
Anther.	*Opening lengthwise, oblong, innate.*
PISTIL—OVARY.	*Triple, ovoid, 6-ovuled, green.*
Style.	*Slender, green-white.*
Stigmas.	*3, pink-colored.*
FRUIT.	*Capsule, 3-sided, conical, inclosed in the calyx.*
Pericarp.	*Dry, 3-valved, 1-celled or partly 3-celled.*
Seeds.	*3—6, lens-shaped, black, shining.*

LOCALITY.—Low, damp soil. (Date), April 25.
CLASSIFICATION.—Subkingdom, **FLOWERING PLANTS.**
—Province, **EXOGENS.**
—Order, PORTULACACEÆ.
NAME.—English, *Spring Beauty.*
—Latin, **Claytonia Virginica.**

ANALYSIS OF 45

ORGANS.	DESCRIPTION.
Plant.	
Root.	
Stem.	
Leaves.	
Inflorescence.	
Flower.	
Calyx.	
Sepals.	
Corolla.	
Petals.	
Stamens.	
Filament.	
Anther.	
Ovary.	
Style.	
Stigma.	
Fruit.	
Seeds.	

LOCALITY, ... DATE, ...
CLASSIFICATION, ... NAME.—Latin, ...
　　　　　Order, ...　　　　　—English, ...

VIII. THE EARLY CROWFOOT.

Description.—In May and June the fields are resplendent with Buttercups. As early as April we find one kind, at least, gilding the rocky hills and woods. In this, and its portrait, root, stems, leaves, flowers, stalks, and fruit are present.

Analysis. — The *Root* is a bundle (*fascicle*) of fibers, some of which are thickened, fleshy, almost tuberous; we call such roots *fasciculate*. They are strong and durable. They have survived the frosts of the past winter; and if you have considered the Crowfoot plant from year to year, you have learned that it is a *perennial herb*. Although the parts above ground perish in Autumn, the root still lives and sends up another plant in the following Spring, and so on for many seasons. The symbol adopted for such an herb is ♃.

FIG. VIII.—Ranúnculus fasciculàris: 3, a head of carpels; 4, a single carpel; 5, the seed in the achenium; 6, seed dissected, showing the embryo in albumen.

The *Stem* (or stems, for there may be several arising from the same root) is green and herbaceous, branching, hairy, 6–10' high, and ending in the flower-stalks.

The *Leaves* are many. Most of them are *radical*, arising with the stem from the root (*radix*); others are *cauline*, growing from the stem (*caulis*) above the base. They are also *alternate* in arrangement—one above another, spirally (p. 193). The lower leaves are borne on long, slender *petioles;* the upper are *sessile*, having no petiole. The blades are *ternately* divided, i. e., into three divisions, or *pinnately* into five, and the terminal division is stalked ; then all the divisions are cut into lobes.

The *Veins*, although present, are not always discernible. In Spring Beauty they are concealed. In Crowfoot they are not conspicuous, yet evidently are not parallel, as in Erythronium. Under a magnifier they seem to form a network ; that is, they are *reticulated* (*reticula*, a net).

The *Flowers* are borne on peduncles, which terminate the stem and branches—one on each peduncle. They are *perfect* and *complete*, having all the proper and essential floral organs.* The calyx consists of five lanceolate, greenish, spreading sepals, much smaller than the petals. The corolla consists of five yellow, shining, spreading petals. In outline the petals are inversely lanceolate or ovate (*ob*-lanceolate or *ob*-ovate), being broadest above the middle. At the base there is a honey-pore, which we may call a *nectary*, covered by a little scale (Fig. IX, 2). The stamens are many in number—said to be *indefinite* (denoted thus, ∞), although, if carefully counted, they will generally be found some multiple of 5; as 25, 30, etc. The term *polyandrous* (*polys*, many, *andres*, stamens) is of a similar meaning. The filament, anther, and pollen we leave to be identified by the student.

---

* A flower is botanically *perfect* when it has both stamens and pistils. The symbol is ⚥. It is *complete* when it has stamens, pistils, calyx and corolla—all the proper organs. A flower is *apetalous* when the calyx is present without the corolla ; it is *staminate* (♂) when having stamens without pistils, and *pistillate* (♀) pistils without stamens.

Notice how the stamens stand directly on the torus, neither *ad*-hering to any other member, nor *co*-hering among themselves. They are *hypogynous* (*hypo*, under, *gynè*, the pistil). This character is of great significance.* (Fig. VIII, 9.) The pistils are also numerous, twenty or more, generally some multiple of 5. Their form and structure are remarkable— *one-sided* (5), consisting each of an ovary tipped with a sessile stigma, without a style.

The *Fruit*. In a few days the work of the yellow buttercup is done.

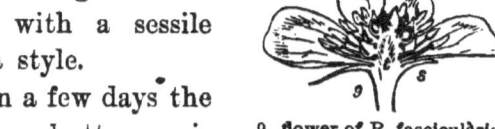

9 flower of R. fascicularis.

Bees and other insects have drained its nectaries and scattered its pollen. The sepals, petals and stamens fade and fall. These are the deciduous parts. But the pistils still persist, attached to the torus, growing and forming a roundish head (4) of as many little fruits (carpels) as there were pistils. Let us dissect one of these carpels (6). It holds just one seed in one cell. It is an *achenium*—a simple fruit formed of one carpel (not of three, as in Erythronium). In the figure is represented a section of the seed, showing a small embryo with two cotyledons, imbedded in albumen. Here is work for the microscope.

The Name.—There are many kinds of Buttercup-Crowfoots. Some of them delight in ponds and sluggish streams, with the frogs for their companions. For this reason, Linnæus named them all Ranúnculus (a little frog). *Ranúnculus* is therefore the name of a group of similar forms, = a Genus, including all sorts and kinds of Buttercup-Crowfoots. The specific form here figured and described, known at sight by its early date,† showy flowers,

---

* On account of their hypogynous stamens, and the entire freedom or *distinctness of all* their floral organs, botanists have assigned the Buttercups and their order to the highest rank in the Vegetable Kingdom.

† There is no danger of confounding this species with that other one which also

RANUNCULUS.

ANALYSIS OF EARLY BUTTERCUP.

ORGAN.	(Its) Life, Habit, Number, Place, Kind, Construction, Form, Size, Qualities of color, surface, taste, &c., and Appendages.
THE PLANT.	♃ *damp shades. 1 foot high. Hairy.*
ROOT.	♃ *fasciculate, fibers white, long, some of them thickened.*
STEM.	*Herbaceous, branching, caulis hollow, diffuse.*
LEAVES.	*Deciduous, alternate, pinnately divided, netted, petiolate.*
INFLORESCENCE.	*Terminal, erect, solitary, peduncle 1–6', terete.*
FLOWER.	*5-parted, complete, regular, 1' broad.*
Calyx.	*Spreading, greenish-yellow.*
Sepals.	*5, deciduous, lanceolate, distinct, imbricate.*
Corolla.	*Rosaceous, shining golden-yellow.*
Petals.	*5, deciduous, oblanceolate, scale and honey-pore at base.*
Stamens.	*30–40, hypogynous, with slender filaments.*
Anthers.	*Oblong, 2-celled, yellow, dehiscing lengthwise.*
Pistils.	*20–30, distinct, style very short or none.*
Ovary.	*Obliquely ovate, lens-shaped, green.*
Stigma.	*Sessile, terminal, a little curved.*
FRUIT.	*20–30, distinct, achenia, in a roundish head.*
SEED.	*One in each carpel or achenium.*

LOCALITY—*Woods, Westchester, N. Y.*

CLASSIFICATION.—**PHENOGAMIA, EXOGENS.**

ORDER.—RANUNCULACEÆ, or THE CROWFOOTS.

NAME.—Latin, **Ranunculus fascicularis.**

—English, *Early Crowfoot. Buttercups.*

begins to flower in April, having its radical leaves nearly round, *crenate*-toothed (see Glossary), its petals very small, not larger than the sepals, and therefore named by Linnæus, *R. abortivus* (as if the flower were abortive), and *fasciculate* roots, was named by Dr. Muhlenburg,* *Ranúnculus fasciculàris.*

Scientific Terms.—Achenium. Alternate. Cauline. Fasciculate. Hypogynous. Indefinite. Nectary. Oblanceolate. Obovate. Perennial Herb. Pinnately divided. Radical. Reticulate. Sessile. Simple fruit. Ternately divided.

IX. THE BULBOUS CROWFOOT.

Description.—In the month of May we begin to find other Crowfoots, especially the Bulbous, adorning the meadows and hilly pastures with their golden cups. This is a neat and elegant plant, more erect and silky-haired than the Early C. Indeed it is pre-eminently the true Buttercup. Let us see wherein the two kinds differ, and how they may be distinguished.

Analysis.—The *Root* is fibrous, being wholly composed of slender, white, branching fibers, springing together from the base of the stem. There is no central axis as if the stem continued downward. Such roots are *inaxial.*†

The *Stem* enlarges at the base into a sort of bulb, rather a *corm*, which in the Autumn is round, plump and solid.‡ Thence it stands erect, about 1 foot high, dividing into straight branches ending in flower-stalks.

---

* Rev. Henry Muhlenburg, D.D., author of a work on the Grasses, Flora Lancastriensis, and other books, was in his day the pioneer American botanist, "a Christian philosopher characterized by zeal and industry not more than by humility and unbounded liberality of sentiment towards his contemporaries." Died A. D. 1815, æt. 62.

† This term will be better understood if we compare it with an *axial* root, such as we find in the Clover or Yellow Dock, where the stem seems to continue downward, gradually dissolving into fibers.

‡ In ancient times this bulb was called "St. Anthony's Turnip." But if that pious hermit ever dined on it, he must have dried it well in the sun to expel its acridity. In its fresh state it is pungent and emetic—properties of which medical students sometimes make a mischievous use by persuading their companions in attendance on the botanical lectures to test their excellence by tasting. The herbage also has acrid properties, which prove a defence against its enemies. Cattle avoid it, so that it stands and blooms unmolested even in closely cropped pastures.

RANUNCULUS BULBOSUS.

Fig. IX.—Ranúnculus bulbôsus: *A*, the bulb, as in autumn; *B*, the bulb in spring; 1, plan of the flower; 2, a petal; 3, achenium dissected.

The *Leaves* are mostly radical, long-petioled, ternately divided, with the terminal division stalked, all deeply 3-cleft, and lobed. The venation is plainly reticulated.

The *Flowers* are singly mounted on long, slender peduncles which are grooved or furrowed. The 5 sepals are *reflexed*—bent backward and downward.* The 5 petals are broad, rounded, shining and golden, forming a cup-shaped corolla. The honey scale at the base of each petal is toothed. The stamens are about 50; and the pistils (carpels) about 20, each tipped with a short, sessile, recurved stigma.

The *Fruit* is a round head of about 20 distinct, lens-shaped *achenia*, each tipped with a short beak.

PLAN OF THE FLOWER.—While there is only one whorl or circle of sepals (*c*) and one of petals (*p*), there are at least 5 of stamens (*s*) and 4 of pistils (*o*). The *alternating* position of all these organs, so clearly shown in the diagram, is obscured in the flower itself by their crowded condition. Why is the flower symmetrical? Why is it hypogynous?

The Name.— This pretty specimen of Buttercup is appropriately named *Ranúnculus bulbòsus* (Linn.) †—the Bulbous Crowfoot.

The Record.—The analysis of this plant may be recorded in the accompanying blank tablet, or in one of similar

---

* It is noticeable that the green sepals of the Calyx, having acted as nurses and protectors to the petals of the flower buds, are reflected or fall off almost immediately after the flower opens, as if they were anxious not to interfere with the success of the floral functions by concealing the bright petals from the insect eye.

† Of the genus Ranunculus there are 50 species in N. America, and at least 200 in the World. Their prevailing color is yellow, but some are white, as the beautiful R. aconitifòlius of the Alps, and the gardens. Another, the splendid R. Asiáticus, is either yellow or crimson on the hills of Palestine. This is the Garden Ranunculus which sports into innumerable varieties of color, with single or double flowers as large as a Rose.

RANUNCULUS BULBOSUS.

ORGAN.	(Its) *Life, Habit, Number, Place, Dehiscence, Kind, Construction, Form, Placentation, Size, Qualities, Appendages.*
Plant, L.H.S.Q.	
Root, L.K.	
Stem, L.H.K.F.	
Leaves, L.P.C.F.S.Q.	
Inflorescence, P.K.A.	
Flower, N.C.	
Calyx, F.Q.	
Sepals, L.N.P.F.	
Corolla, F.Q.	
Petals, L.N.P.F.	
Stamens, N.P.C.	
Anther, D.C.F.	
Style, N.C.F.	
Stigma, N.F.	
Ovary, C.F.Pn.	
Fruit, N.D.K.F.Q.	
Seed, N.C.F.Q.A.	

LOCALITY,... DATE,.................

CLASSIFICATION,................ | NAME.—Latin,................

—Order,................ | —English,................

construction. The letters following the name of the organ are the initials of topics at the head of the tablet.

Scientific Terms.—Axial root. Inaxial root. Reflexed.

X. THE LIVERLEAF.

Description.— In the rich, black mould of the rocky woods, low among the dead leaves where a snowbank lately lingered, peeps up the blue Liverleaf to herald, with the Bluebird, the advent of Spring. Nothing is visible but its leaves and flowers. When lifted from its bed entire and its roots rinsed clear of soil, the plant may be placed in a glass of water, and examined at leisure.

Analysis.— The *Root* is inaxial like that of Ranunculus, consisting wholly of long branching fibers.

The *Stem* is a shapeless body—the solid basis of the plant under ground, whence spring the roots downward, and the leaf and flower stalks upward. Such a stem is called the *crown*, and the plant is said to be *acaulescent* (stemless), for it has no proper stem. Plants with ordinary stems bearing the leaves and flowers, like the Crowfoots, are said to be *caulescent*.

The *Leaves* are of two ages, some of the last year's growth, and some just arisen. All are supported on long, hairy stalks or petioles which arise from the ground, apparently from the roots; hence they are *radical* and *petiolate*. The blade is firm and leathery in texture, that is, *coriaceous*, and is fashioned into 3 entire lobes (*trilobate*). In the portrait (Fig. X) the lobes of the leaves are *obtuse* at the apex. Is it so in your specimen?

The VENATION of the blade is hand-shaped or *palmate*. From the end of the petiole in the base of the blade, 3 primary veins diverge, one through each lobe to its apex.

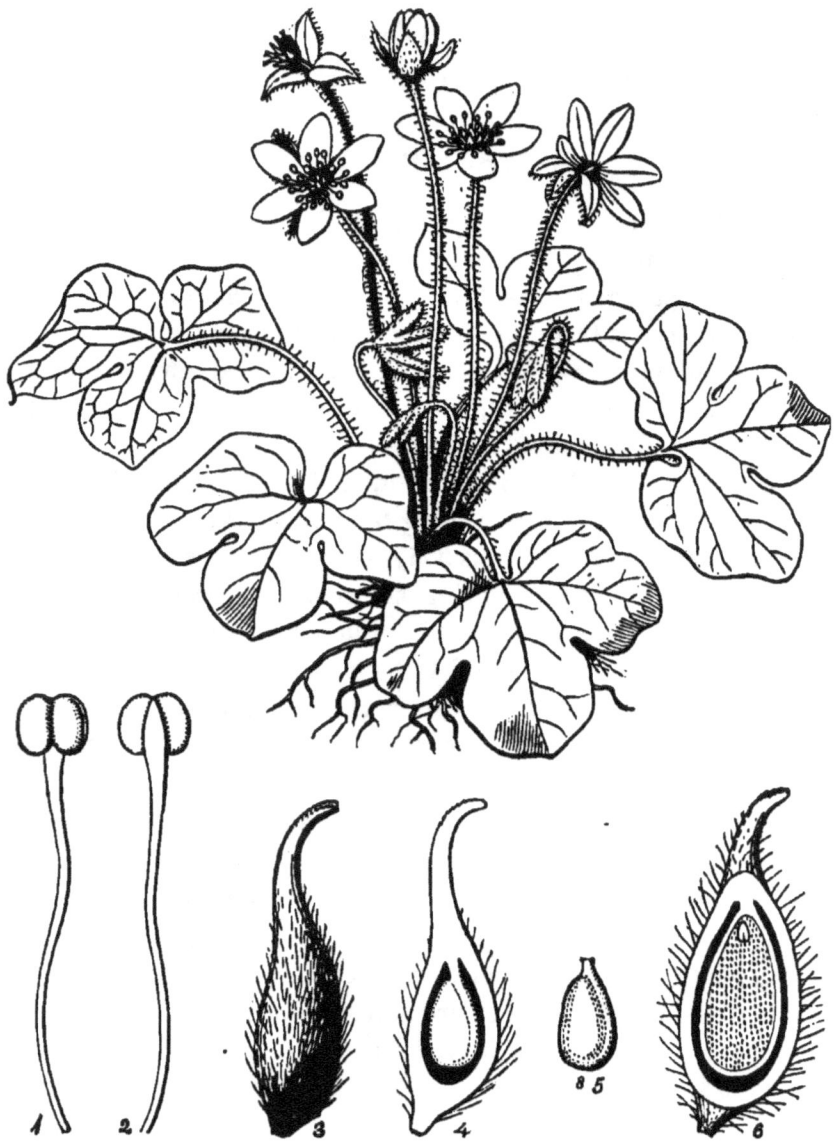

FIG. X.—Hepatica (or Anemône) triloba : 1, a stamen seen in front ; 2, — in rear ; 3, a pistil or carpel ; 4, 5, the ovule, pendulous, anatropous ; 6, a section of the seed showing the 2-cotyledoned embryo in large albumen.

From these veins along each side proceed the *veinlets*, and from the veinlets start out the *veinulets*. The latter form a fine network throughout, and so differ from the forked veinulets of the Ferns (p. 22). Hence this leaf is *palmi-veined* and *net-veined*. Take note also of the persistence of the leaves. They survive the frosts and snows of Winter until after the new leaves of the following Spring appear. Thus the plant is *evergreen*.

Inflorescence. The flowers are mounted each on a slender stalk arising from the crown with the petioles. Such stalks, bearing no true leaves, but flowers only, are called *scapes*. Near the top of the scape, a little below the flower, is a whorl of 3 little green leaves egg-shaped or *ovate** in outline. Are these the sepals of the calyx? They are not so regarded, for they are remote from the flower. They are *bracts* forming an *involucre* (*involvo*, I wrap up).

The calyx resembles a corolla. There are 6 to 9 colored, oblong or obovate sepals, white or delicately tinged with blue or purple. But why is this circlet called a calyx rather than a corolla? It is so named in accordance with a general rule that "the outer whorl of the floral envelopes shall be regarded as the calyx whatever be its color." Hence the corolla is wanting in this flower, as there is no second interior whorl, and the flower is *apetalous* (without petals). The stamens, pistils and fruit are so nearly like those of Ranunculus that the student will need no repetition of the terms to be employed in their description.

The Ovule (4, 5) grows out or is *pendulous* from the top of its cell. The stalk (funiculus) passes down its side to *s*, or rather say the ovule is *anátropous*, i. e., turned or bent over on its stalk. How is it in Claytonia (p. 42)?

---

* The term *ovate* is employed in describing flat, expanded bodies, like leaves; *ovoid* is applicable to solids, such as the bulb.

HEPATICA.

Classification.—The plant represented in Fig. X, as named in our botanies generally, is *Hepatica tríloba* (*hepar*, liver, *triloba*, 3-lobed). The early Linnæan name (probably the true one), was Anemòne Hepática. Our specimens may be of the other species, H. acutíloba. In the former the leaf-lobes and bracts are *obtuse*; in the latter, *acute*. As we have seen, Hepática is closely related to Ranúnculus, especially in

ORGAN.	Life, Habit, Number, Place, Kind, Construction, &c.
PLANT.	♃, *acaulescent herb, 3—6′ high.*
ROOT.	♃, *of many long branching fibers.*
STEM.	*Crown subterranean, perennial.* [*venation palmate.*
LEAVES.	*Evergreen, coriaceous, trilobate, acute on radical petioles,*
INFLORESCENCE.	*Scapes radical, 1-flowered, pubescent.*
FLOWER.	*Apetalous, with an involucre of 3 ovate bracts.*
Calyx.	*Corolla-like, light blue or purple.*
Sepals.	*6—9, oblong or obovate.*
Corolla.	*Wanting.*
Petals.	*Wanting.*
Stamens.	∞, *hypogynous, white, filaments slender.*
Anthers.	*Oblong, 2-celled.*
Pistils.	∞, *green, hairy.*
Ovary.	*Oblong, distinct, simple.*
Stigma.	*Nearly sessile, acute, style none.*
FRUIT.	*12 or more oblong achenia hairy at top.*
SEED.	*One in each carpel.*

LOCALITY.—*Dry woods.* (Date), *April 10, 1877.*
CLASSIFICATION.—**PHENOGAMIA, EXOGENS.**
Order, RANUNCULACEÆ.
NAME, **Hepatica acutiloba.**

the absolute freedom or distinctness of all its organs, and the structure of the stamens, pistils and fruit. It must therefore be included in the same Tribe or Order, viz., Ranunculaceæ, or the Crowfoots.

Scientific Terms.— Acaulescent. Apetalous. Bracts. Caulescent. Coriaceous. Evergreen. Involucre. Palmi-veined. Scape. Trilobate. Veinlets. Veins. Veinulets.

XI. THE RUE ANEMONE.

Description.— In April and May the woods, while yet leafless, are aglow with Anémones. The species portrayed in Fig. XI continues long in bloom, developing its pure white flowers in succession until a full cluster is displayed.

Analysis.—The *Root* is similar to that of Claytonia (p. 39), but instead of one there are usually 2 or 3 tubers of an oblong form, with fibrous roots attached. As to its life or duration, it would seem to be perennial (♃).

The *Stem*—its life, habit of posture and branching, its form and dimensions (size) may be considered and noted by the student.

The *Leaves* are *compound*, and will furnish the principal topic in this analysis. There is one radical leaf and 2 or 3 cauline. The former is also petiolate. The petiole divides at the top into 3 branches (*petiolules*) and these again into 3s—9 in all, each bearing a *leaflet*. It is therefore *twice ternate* or *biternate*. The cauline leaves are situated at the top of the stem (*inv*). Apparently there are 6 or 9 simple, petiolate leaves in a whorl. But the petioles are joined at the base into sets—3 in each set. Hence we conclude that there are 2 or 3 *ternate* or *trifoliate*, sessile leaves. The leaflets are all similar, *oval*, 3-lobed at the end. Not unfrequently the radical leaves are thrice 3-parted, bearing 27 leaflets. They then become *triternate*.

THE RUE ANEMONE. 59

The *Inflorescence* is terminal. The leaves around it may be regarded as forming its involucre. Several pedicels, each bearing a flower, arise from a common point in the midst of the leaves. Such an arrangement is called an *umbel* (*umbella*, a little shadow), and the pedicels are the *rays* of the umbel.

The *Flowers* repeat the apetalous habit of Hepatica. There is a single whorl of envelopes— the calyx, composed of 5 to 10 distinct, elliptical sepals of dazzling white. There is a crowd of stamens, with side-opening anthers, perfectly distinct and free; and in their midst appear the 6–10 distinct pistils. The close observer will miss the nectaries.* Neither

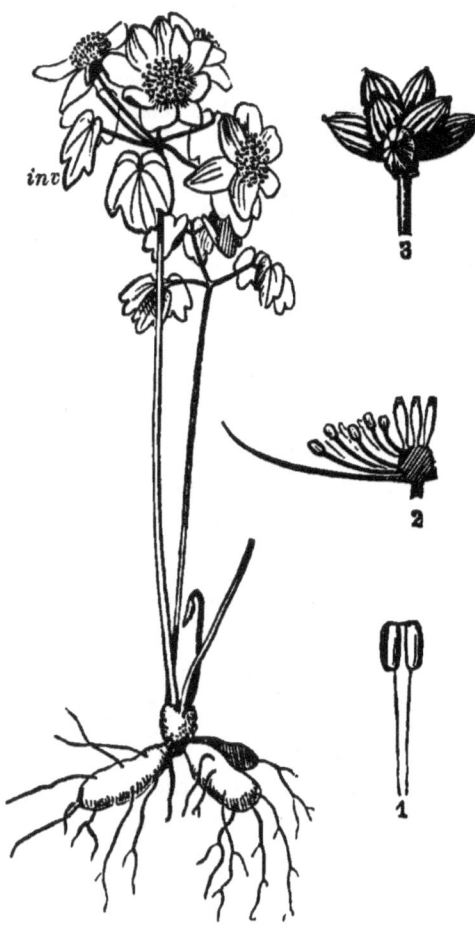

Fig. XI.—Anemòne thalictroìdes: 1, a stamen; 2, section of flower; 3, fruit.

---

* The Ranunculaceæ offer very remarkable differences in the manner of their adaptation to insects. Honey is secreted by the petals in Ranunculus, Hellebore, Larkspur and Columbine; by the sepals in certain Pæonies, by the stamens in Pulsatilla, and by the ovaries in Cowslip, while it is entirely absent in Anemone, Hepatica, and Thalictrum. The flower is made conspicuous by the corolla in Ranunculus, by the calyx in Anemone, by both in Larkspur, and by the stamens in Thalictrum. The

honey scales, nor glands, nor spurs are to be found in this flower, which is *regular, apetalous, polyandrous, and hypogynous.*

The *Fruit*. After the sepals and stamens have perished, the green pistils still persist and ripen into a head of distinct achenia which are singularly grooved and fluted.

XII. THE WOOD ANEMONE.

"The coy Anémoné that ne'er uncloses
Her lips until they're blown on by the Wind."

Description.—The Wind Flower, as it is frequently called, abounds in hilly woods and often in company with the Rue Anemone. It is a smaller plant, always one-flowered, and about 5' high while the latter may be 9'.

Analysis.—In the *Root* we have a new feature. It is a slender creeper, a little fleshy, growing just beneath the surface of the soil. It is called the root-stock, or more accurately the *rhizome*. From its joints fibers grow downward and stems upward.

The *Stem*, slender but firm and erect, bears at the top 3 compound leaves forming, as it were, an involucre around the one large flower. There is often, also, a radical leaf of the same form. All are petiolate, palmately compound, and their 3 (rarely 5) leaflets wedge-shaped (*cuneate*) at the base, cut into lobes and teeth above. They are acrid to the taste like the herbage of the Buttercup. Sheep and goats will eat them, however, while they are refused by cattle and swine.

---

honey is easily accessible in Ranunculus to all kinds of insects, yet the flower can dispense with their services and fertilize itself; while in Larkspur, where insect aid is indispensable, the honey is stowed away in the end of deep spurs, and accessible to bees only. The stigmas are not matured until after their own stamens have shed their pollen; then they put themselves in the way of the bees, to be dusted with pollen from other flowers.

The *Inflorescence* is solitary. The one large flower is near of kin to the foregoing. It is apetalous. Its 4—7 sepals are oval in outline, white and more or less tinged with purple. The stamens and pistils will also be identified and defined by the student. Why are they indefinite? Which hypogynous?

The *Fruit*. Is it compound, or simple? Of what kind? How many seeds in each little fruit or carpel?

The Name.—*Anemòne*,* the generic title, comes from the Gr. *animos*, wind. It was adopted by Linnæus from the idea then prevalent that its flowers open only when the wind is blowing. The specific name of the Rue Anémone is *A. thalictroïdes*, so called for its resemblance to Thalictrum, the Meadow Rue.† Of the Wood

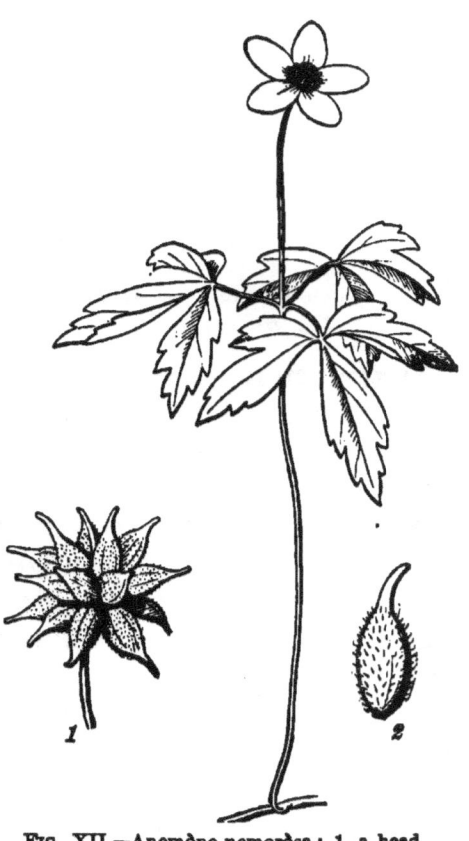

FIG. XII.—Anemòne nemoròsa: 1, a head of ripe carpels; 2, a single carpel—achenium.

---

* So accented according to the Latin prosody; but as an English word it is Anémone.

† This plant has long hung trembling between the two genera Anemone and Thalictrum. Its involucre and flowers are those of Anemone; its leaflets and achenia like those of Thalictrum. Linnæus named it as above. Michaux called it Thalictrum anemonoides.

Anémone, *A. nemoròsa* (*nemus*, a grove) is the specific name.*

In the Meadow Rue, the minute sepals fall off as soon as the flower opens. But the stamens are enlarged and their anthers yellow. Thus a little floral economy does away with the necessity of the usual attractive floral envelopes.

Classification.—The next inquiry is, To what order do the Anemonies and Hepaticas belong? With stamens polyandrous, hypógynous, and pistils distinct, forming simple, unconnected fruits, they agree with the Crowfoots, and their order is RANUNCULACEÆ.

The Order of the Crowfoots (Ranunculaceæ) embraces in all about 55 genera and 1100 species. From the foregoing and other examples, we deduce the following brief formula of its character:

1. Plants with a colorless, acrid juice.
2. Leaves reticulate-veined, never peltate.
3. Flowers with their members all free and distinct.
4. Sepals, or petals, 3—15, equal or unequal.
5. Stamens indefinite, hypogynous.
6. Pistils few or many, distinct, oblique.
7. Fruit a few or many achenia, pods, or berries.

The Crowfoots delight in cool, damp climates. Their juices, generally acrid, are strong enough in some Butter-

---

* The genus Anemone is large and interesting. Sixty kinds inhabit the N. Temperate Zone in both Continents. In the United States, from Sea to Sea, some 20 species flourish. The *Pasque Flower*, Nuttall's Pulsatilla, in Illinois and the Northwest, is the most curious of them all. Its bluish blossoms, as large as a Rose, open in early April; after them the leaves, cut into many slender segments and clothed with long silky hairs, spread themselves, while the ripened achenia, fledged with feathery tails, take flight on the wind to new and distant homes. It derives its name from the fact that it was formerly the custom in England to use this, as well as other wild flowers in staining Egg for Easter-gifts, called *Pasque Eggs*.

Many grades and styles of beauty are represented in this genus, from the humble Wood Anemone upward, culminating in the Royal Anemone of Palestine (A. coronària), one of the "Lilies of the field" arrayed in more than Solomon's glory. Its leaves are delicately cut and fringed, and its flowers, broad as the hand, shine in Tyrian purple,

ANALYSIS OF AN ANEMONE.

ORGAN.	Life, Habit, Number, Place, Dehiscence, Kind, Construction, Form, Placentation, Size, Qualities, Appendages.
Plant, L.H.S.Q.	♃, *herb 6–10′ high, generally smooth.*
Root, L K.	♃, *oblong starchy tubers with fibers attached.*
Stem, L.H.K.F.	*Annual, an erect, simple, terete caulis.*
Leaves, L.P.C.F.S.Q.	*1 radical, biternate, 2 cauline, ternate, sessile, lfts. 3-lobed.*
Inflorescence, P.K.A.	*Terminal, umbellate, involucrate.*
Flower, N.C.	*3–7, regular, apetalous, hypogynous.*
Calyx, F.Q.	*Rose-form, petaloid, white.*
Sepals, L.N.P.F.	*Deciduous, 5–10, spreading, elliptical, imbricated.*
Corolla. F.Q.	*None.*
Petals, L.N.P.F.	*None.*
Stamens, N.P.C.	*∞, hypogynous, distinct, filament club-shaped (clavate).*
Anther, D.C.F.	*Oval, 2-lobed, opening laterally, innate.*
Style, N.C.F.	*None, or very short.*
Stigma, N.F.	*6–10, sessile, simple.*
Ovary, C.F.Pn.	*Distinct, simple, oblong, with 1 suspended ovule.*
Fruit, N.D.K.F.Q.	*Achenia 6–10, smooth, fluted, ovoid.*
Seed, N.C.F.Q.A.	*1 in each carpel, albuminous, emb. 2-lobed.*

LOCALITY.—*Woods, Worcester, Mass.* (Date), *May 3, 1870.*
CLASSIFICATION.—**PHENOGAMIA; EXOGENS.**
    ORDER.—RANUNCULACEÆ, or the CROWFOOTS.
        NAME.—Latin, **Anemone thalictroides.**
           —English, *Rue Anémoné.*
REMARKS.—*The cauline leaves serve as an involucre.*

cups to blister the skin, and become actively medicinal and poisonous in Aconite and Hellebore. Their flowers are generally ornamental, of various styles of beauty in Clématis, Adònis, Ranúnculus, Anémone, Columbine, Larkspur, Monk's-hood, and culminating in the splendid Pæony.

The Record.— Let the student now enter in the tablets of the Plant Record, or such as he may himself prepare, the analysis of the Anemonies. In doing it, the presence of the plant itself is indispensable, together with the foregoing instructions, and also a frequent reference to the Illustrated Glossary.

Scientific Terms (defined in XI and XII).—Biternate. Compound leaves. Cuneate. Generic. Leaflet. Palmate. Petiolule. Rays. Rhizome. Specific. Ternate. Trifoliate. Triternate. Umbel.

XIII. THE BLOODROOT.

Description. — Some sunny morning in Spring, in woody vales along the banks of a purling brook, or the track of a hidden streamlet, we may surprise in bloom the bright, frail flowers of the Bloodroot. The plant is remarkably simple in its portrait, smooth and glaucous in surface.

Analysis.—The *Root* consists of fibers and fibrils only, for we must consider that thick, fleshy body (*rh*), although underground,

The *Stem;* there is no other. It is a true rhizome or root-stock, growing horizontally, filled and reeking with a blood-red, acrid, medicinal juice.* From its joints or off-

---

* In lifting this plant from its bed, one is forcibly reminded of the sad experience of Æneas at the grave of Polydore (Æneid, Book III).

I pulled a plant—with horror I relate
A prodigy so strange, and full of fate !
The rooted fibers rose, and from the wound
Black bloody drops distilled upon the ground.

SANGUINARIA.

Fig. XIII.—Sanguinària Canadénsis*: 1, 2, stamens; 3, the pistil; 4, pistil dissected; 5, an ovule anatropous; 6, the capsule opening; 7, seed; 8, seed dissected, both crested; 9, the embryo.

* Generic names are nouns, and should always begin with a Capital. Specific names are generally adjectives, and should never begin with a capital except when derived from (1) a country, or (2) a person, or (3) when they are nouns; as (1) *Sanguinaria Canadensis*, (2) *Osmunda Claytoniana*, (3) *Papaver Rheas*.

sets here and there, a bud issues and sends up a leaf and a flower—the whole visible plant. The term *acaulescent* is applicable to the Bloodroot as well as to Hepatica, it being apparently stemless.

The *Leaf* comes up from the ground tenderly embracing the flower-bud like a cloak. In the bud both together were enveloped in the membranous scales (*sc*) which now surround the base of the stalks. The rounded blade is conspicuously palmi-veined and netted, its margin lobed, with rounded *sinuses* between the lobes, and its base *cordate* (heart-shaped) with a deep recess.

The *Flower* loses its two green *caducous* sepals as soon as it opens. The pure-white petals, open only in the sunshine, are soon deciduous. The 4 interior are shorter than the 4 exterior, giving the expanded corolla the form of a square. The stamens, about 24 in number, are hypógynous.

The *Fruit.* A pistil evidently composed of 2 united carpels, having a double, sessile stigma (3, 4) occupies the center of the flower. It becomes in fruit an oblong capsule with a single cell. Two lines run lengthwise on opposite sides of the ovary or pod (3) marking the conjoined edges of the carpels. We will call these the *sutures* (*sutura*, a seam). Within the cell are two corresponding lines or ridges to which the seeds are attached; term these the *placentæ;* and being located on the wall (*paries*) of the cell, *parietal placentæ.** It contains many seeds, and finally opens by 2 equal valves which break away from the placentæ and leave them still in place (6). See how the ovule (5) is bent over and adhering to its stalk (*anátropous*). Notice in the seed (7) its prominent and singular crest (*c*) and the 2-lobed (*dicotylèdonous*) embryo (8, 9) in the end of the large albumen.

---

* Observe that in Erythrònium the placentæ are *central.*

SANGUINARIA.

The Name.—This genus is appropriately named *Sanguinària* (*sanguis*, blood); and as Linnæus obtained his first specimen from Canada, he made its specific name *S. Canadénsis*. It is the only species yet known.

Classification.—We find in this plant a new combination of characters—*a colored juice, a flower 2 or 4-parted, a*

ANALYSIS OF BLOODROOT.

ORGAN.	Life, Habit, Number, Place, Dehiscence, Kind, &c.
PLANT.	♃, *acaulescent, 6—10' high, smooth, glaucous.*
Root.	*Fibrous, growing from the thick root-stock.*
Stem.	*A rhizome, horizontal, full of a red acrid juice.*
Leaves.	*Solitary, radical, palmately 7–9-lobed, petiolate.*
Inflorescence.	*Solitary, radical, bud infolded by the leaf.*
Flower.	*One, on a scape, white, perfect, complete.*
Calyx.	*Green, very smooth.*
Sepals.	*Caducous, 2, imbricate in bud.*
Corolla.	*Square in outline, white, double.*
Petals.	*8, in 2 rows, oblanceolate, wide-spread.*
Stamens.	*24, hypogynous, with slender filaments.*
Anther.	*Innate, oblong, yellow.*
Pistil.	*Double, of 2 carpels, green.*
Stigmas.	*2, sessile or style very short.*
Ovary.	*Tapering at both ends, ovules in 2 rows.*
Fruit.	*Capsule turgid, 1' long, 1-celled, 2-valved.*
Seeds.	∞, *round, crested on one side, reddish brown.*

LOCALITY—*Damp woody vales.* (Date), *April 10, 1877.*
CLASSIFICATION.—PHENOGAMIA, POLYPETALOUS EXOGENS.
Order, PAPAVERACEÆ POPPYWORTS.
NAME.—Latin, **Sanguinaria Canadensis**.

caducous calyx, numerous hypogynous stamens, and a compound 1-celled capsule. Let these be remembered as the marks of the Order, which will be named in the next lesson.

The Record (on page 67) should be used by the learner not to copy, but for comparison with his own, previously and independently sketched; also as a guide in the record of the Poppy and other similar plants.

XIV. THE POPPY.

Description.—Toward the end of May some of the Poppies may be found in bloom in gardens and fields. Their graceful form attracts the eye, while the richness of their scarlet tint harmonizes with the green verdure around. Their own verdure is sea-green, somewhat hairy, and like the Bloodroot, contains a colored juice—white instead of red. The Poppy never springs from the last year's root, but from the seed alone, flowering, fruiting, and perishing, all in one season. It is therefore an *annual herb* (often thus denoted ①). The Bloodroot with its ever-growing rhizome is necessarily perennial (♃).

Analysis.—The root is *axial* (p. 50)—a tap-root growing from the seed downward, branching, tapering.

The *Stem* stands firmly erect, terete, somewhat branching, and with bristly spreading hairs.

The *Leaves* are cauline, sessile, pinni-veined, and oblong in general outline, with the margin more or less lobed, or divided into segments (*pinnatifid*).

The *Flowers* are few and large, each supported on a stout peduncle, nodding in the bud (*b*), finally erect. The

---

* These notes apply only to the natural, *single*, or *simple*-flowered Poppy. Should the specimens have *double* flowers, they will open a new field of inquiry, for which see the lesson on the Rose (XXVIII).

FIG. XIV.—Papàver Rheas: 1, the fruit.

calyx is like that of Sanguinaria. The corolla (white or red) consists of 4 broad petals, thin and fragile, crumpled in the bud and opening convulsively.* The stamens are as in Sanguinaria, but more numerous. There is but one pistil, a large, turgid, green ovary capped by a broad, sessile stigma, with no style. The *rays* marking the top of the stigma, indicate so many simple stigmas and carpels united into this one compound pistil.

The *Fruit* is a capsule, 1-celled, crowned with the broad, persistent stigma. It opens by as many little valves under the margin of the stigma as it has rays—one to each carpel, for the escape of the seeds. These are exceedingly small and numerous. Linnæus counted 10,000 in a single capsule.

The Name of the Poppy family is the ancient Roman one, *Papáver*. It is said to come from *papa*, the Celtic word for pap, because its capsules were formerly given to infants with their food as a soporific. Among the 30 species of the Poppy, red is the prevailing color. Three kinds, at least, with large, brilliant scarlet or crimson petals frequent our gardens and fields. One of these, *P. Rheas*, the Corn Poppy, is portrayed in Fig. XIV.† Another species, the Opium Poppy, has white flowers. It is appropriately named *P. somníferum* (*somnus*, sleep, *fero*, I bear).

The Order or tribe of the Poppyworts—PAPAVERACEÆ,

---

* These petals are so delicate that even when we cut them with scissors it is almost impossible to keep them from crumpling. But the Poppy bee having dug a hole three inches deep in the ground and smoothed and polished the sides, hangs the walls of its little home with tapestry, using these Poppy petals, which it employs with so much skill that they are smooth as glass.

† The Ancients believed that the presence of the Corn Poppy in their fields was necessary to the prosperity of the Corn (Wheat); hence the seeds were among the sacred offerings to Ceres, and her garlands were composed of Wheat-stalks with their bearded heads intertwined with Red Poppies. "The term rheas," says William Turner, who wrote in 1551, "is given because the flower fallith awaie hastilie." This Poppy is so abundant in England that it is dreaded by the farmers as a pestiferous weed.

includes 24 genera and 290 species, chiefly natives of the N. Temperate Zone, briefly characterized as follows :

> Herbs with colored or milky juice.
> Flowers 2 or 4-parted, polyandrous, hypogynous.
> Sepals fugacious.
> Ovary compound, 1-celled.
> Fruit a dry pod, with parietal placentæ.

The POPPYWORTS all possess narcotic properties in their juice, but not in their seeds. The milky white juice of the Opium Poppy, when extracted and dried, becomes the opium of commerce.*

The California Poppy (*Eschscholtzia*), a showy garden annual, is so abundant on the hillsides of California as to paint them with its own yellow-orange color visible far out on the Sea.

The Celandine (*Chelidònium majus*), from Europe, grows in roadsides and waste places. Its saffron-colored juice is said, when faithfully applied, to kill warts.

Scientific Terms.—Annual herb. Caducous. Crest. Dicotyledonous. Glaucous. Placentæ central. Placentæ parietal. Rhizome. Sinus. Sutures. Tap-root.

XV. THE VIOLETS.

Description.—Who does not know and love the Violets? Early or late in spring, in all our rambles, they greet us with their quaint and cheerful faces—the yellow in the rocky woods, the white in boggy swamps, and the blue everywhere.† With specimens in profusion, let us first

---

* The narcotic properties of the Poppy must have been early known, for in ancient Greece the god of sleep was figured as reclining on a bed of its snowy blossoms, and grasping them in his motionless hand. In the East the Poppy attains a greater luxuriance, and its white juice is more abundant than in our colder climate. The process of collecting the opium to-day is the same as described by Dioscorides many centuries ago. At sunset incisions are made in the half-ripened capsules. During the night the juice exudes and collects in globules outside. The next morning these are scraped off, thickened in the sun, and shaped by the hand into balls. The seeds are not injured by the flow of the juice, and make a second harvest. They contain no opium, but are rich in oil, which, as an article of diet, is nearly as good as the Oil of Olives.

† History tells us how in all ages the Violet has been prized. Athens honored it with the first place in floral wreaths. An ancient poet speaks of "living in Violet-

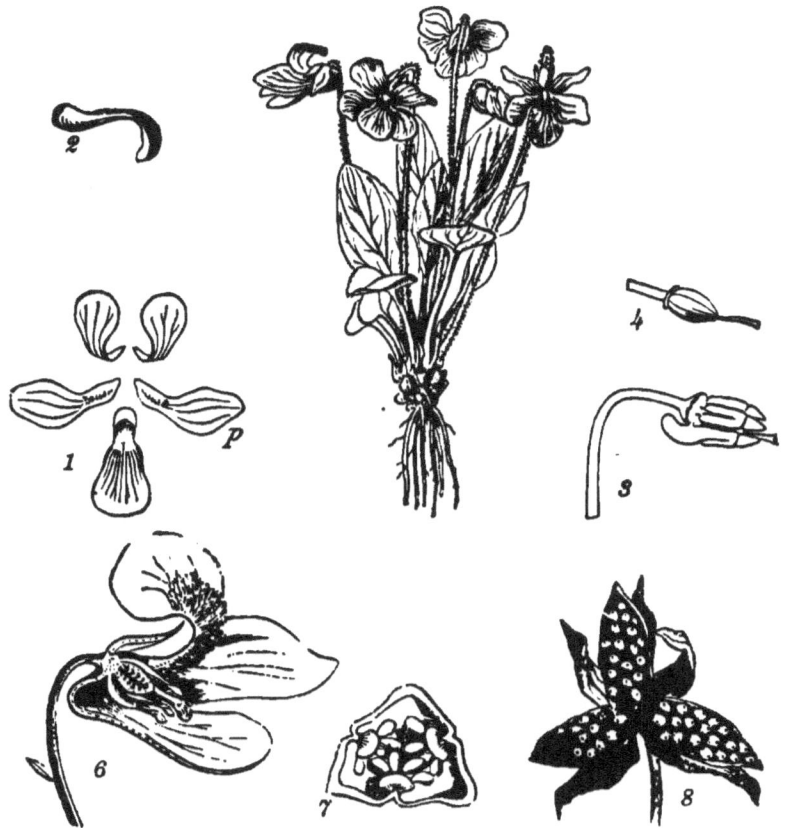

Fig. XV.—Viola cucullàta: 1, the corolla displayed; 2, the odd petal; 3, the stamens, &c.; 4, the ovary and style; 5, section of a seed; 6, section of a flower of V. rotundifòlia; 7, section of ovary; 8, the capsule wide open, the valves covered with seeds.

---

crowned Athens." The Romans had their "Dies Violaris,"—the day devoted to decking the graves with flowers. An old English herbalist says, "Verie manie of these violets receive ornament and comelie grace, for there be made of them garlands for the head, nosegaies and posies, which.... stirre up a man to that which is comely and honest." Who cannot respond to these lines of Willis:

"There is to me
A daintiness about these early flowers
That touches me like poetry. They bloom
With such a simple loveliness among
The common herbs of pasture, and breathe out
Their loves so unobtrusively, like hearts
Whose beatings are too gentle for this world."

examine the blue. Like Hepatica, this Violet is a perennial, acaulescent herb. According to its locality, it is smooth (*glabrous*) or hairy, the latter in poorer soil.

Analysis.—From what the learner has already seen, he will easily characterize the root, stem, and scapes. But in the leaf and flower several new features will appear.

The *Leaves* are borne on long petioles, springing from the underground stem, and each petiole is embraced at its base by a pair of narrow linear scales. The nature of these appendages will be better understood hereafter (p. 75). The blade is rolled inward at the base, so as to imitate the form of a hood (*cucullus*); hence the leaf is *cucullate*. When spread out, as in dried specimens, the blade is as broad or broader than long. A sinus or recess at the base, where it joins the petiole, makes it *cordate* (heart-shaped) or *reniform* (kidney-shaped). The margins are usually *crenate* (notched)—i. e., wrought into small rounded notches. But in this they greatly vary, being sometimes found divided, more or less deeply, in five to nine lobes. As to venation, are they pinni-veined, or palmi-veined?

Inflorescence.—The flower-stalks or scapes are two-bracted about midway, and recurved at the top, so that the flowers are nodding, and *resupinate* (inverted).

The *Flowers* hitherto studied are *regular;* that is, they have the same form and look on every side. But the flowers of the Violet are *irregular*, being oblique or one-sided. This is due to the inequality of the five petals. They differ in shape, size, color, and posture, and are assorted into two pairs and an odd one—the upper (lower by resupination), which is protruded behind into a blunt sack or *spur* (2). All are blue, with a yellow and pencilled base, and the lateral ones are broadest and bearded. The five green sepals are each extended behind into an ear-shaped lobe; that is, they are *auriculate* (*auricula*, a little ear). The five stamens are oddly constructed (3). Hitherto we have seen the anther as in the Crowfoots, *innate*, i. e., borne on the top of the filament; but here it is *adnate*, i. e., attached to the side of the filament below the top. Then two of the filaments project a little spur into the spur of the odd petal !

The *Fruit*. The club-shaped style bears an oblique, hood-shaped stigma. The 1-celled ovary ripens into a 1-celled, 3-valved capsule with three parietal placentæ. (See page 66.) When the valves open they display each a placenta along its middle covered with seeds. Why is the seed (5) anatropous? Why dicotyledonous? What is the ratio of the albumen and the embryo?

9, Cleistogene flowers of V. cucullàta.

Cleistogene Flowers.—The early flowers just described seem to be intended chiefly for display, as they often prove infertile. Later in the season the plant produces flowers on very short scapes, hidden beneath the leaves, or even in the soil, destitute of petals, but always fertile (See (9), where *a* is a flower, *b* a fruit). Such flowers are *cleistogene* (never open), and it is remarkable that their anthers produce but few grains of pollen, barely one to each ovule.*

---

* Here is illustrated the economy of Nature, at one time lavish, at another frugal, but always for a reason. When the pollen is to be carried by chance insects, or perhaps by the wind to distant flowers, an immense amount must needs be wasted. But when it is confined in the closed flower, a very little answers the purpose. In this case there is no need of insect help, and consequently the flowers have no tall stem to push them out into notice, no fragrance, no color, no honey, and indeed no petals. Yet they often bear more seeds than the so-called flowers. There is something almost human in the self-sacrifice of these flowers to sheer duty.

XVI. THE GARDEN VIOLET.

Description.—That the Garden Violet has long been a general favorite is shown by the variety of names it bears, such as Pansy (Fr. *pensée*, thought), Tricolor, Heartsease, Lady's-delight. We find it not only in gardens, but growing wild in fields and woods.*

Analysis.—The *Root* growing downward, branches into innumerable thread-like fibers, which are annually renewed, while as a whole it is perennial.

The *Stem* arises above ground 6–12′, is angular, generally glabrous, with the branches curved upward, leafy throughout.

The *Leaves* are pinni-veined, ovate or oval, crenate, obtuse, and the petiole bears at its base a pair of conspicuous appendages, much larger than those in the Blue Violet

Fig. XVI.—Viola tricolor: 2, a leaf and stipules displayed.

(p. 71), although of the same nature. These are called *stipules*, an organ which distinguishes all the Violets, and many other families; also some whole tribes, as the Rose-

---

* We once (A. D. 1866) crossed a broad plain in central Oregon literally covered with wild Pansies.

ANALYSIS OF A VIOLET.

ORGAN.	Life, Habit, Number, Place, Dehiscence, Kind, &c.
Plant, L.H.S.Q.	♃ *herb acaulescent, 6-12′, smooth.*
Root, L.K.	♃ *an axis or root-stock branching into fibers.*
Stem, L.H.K.F.	*A subterranean crown or rhizome.*
Leaves, L.P.C.F.S.Q.	☉ *on long, radical petioles, cucullate, cordate, palmi-veined, crenate, stipulate.*
Inflorescence, P.K.A.	*Scapes each 1-flowered, with 2 bracts.*
Flower, N.C.P.	*Perfect, complete, nodding, some apetalous.*
Calyx, F.Q.	*Irregular, green.*
Sepals, L.N.P.F.	*Persistent, 5, lance-ovate, auriculate.*
Corolla, F.Q.	*Irregular, chiefly violet-blue.*
Petals, L.N.P.F.	*Deciduous, 5, imbricated, 2-bearded, 1-spurred.*
Stamens, N.P.C.	*5, hypogynous, with short broad filaments.*
Anther, D.C.F.	*Adnate, introrse.*
Style, N.C.F.	*1, oblique, club-shaped.*
Stigma, N.F.	*1, turned to one side, with a beak.*
Ovary, C.F.Pn.	*Triple, ovoid, 1-celled, parietal.*
Fruit, N.D.K.F.Q.	*1, open by 3 valves, capsule, ovoid, smooth.*
Seed, N.C.F.Q.A.	∞, *anatropous, ovoid, brown, appendaged.*

LOCALITY.—*Fields, meadows.* (Date), *April 20, 1877.*
CLASSIFICATION.—PHENOGAMIA, POLYPETALOUS EXOGENS.
    Order, VIOLACEÆ.
  NAME.—Latin, **Viola cucullata.**
    —English, *Hood-leaved Violet.*

worts. Stipules always appear in pairs and attached to the base of the petiole. In shape they are as various as the leaves. Those of the Pansy are cleft into several segments, of which the terminal is the largest—a form called *lyrate*, or *lyrate-pinnatifid*. Contrast these with the stipules of the Blue Violet.

The *Flowers,* in their garden dress, are a perpetual charm, sporting into varieties infinite, yet always with "method in their madness." In their wild or ordinary state, they are definitely tri-colored,* with one petal yellow, two white, and two of that peculiar deep rich purple known as violet, one of the tints of the rainbow. By cultivation the petals may be enlarged tenfold, and their three native colors strangely mixed and confounded. Comparing this flower with the Blue Violet, why is it resupinate? Why irregular? Which petal is spurred, the lower or upper? Which are violet-colored? Why are the sepals auriculate? Which stamens are spurred? Why are the anthers adnate? Being caulescent, this Violet develops no *cleistogene* flowers.

The Name.—*Viola*, the ancient Latin name of these plants,† is adopted in modern science as that of the genus. It includes 150 species. The Blue Violet is *V. cucullàta;* the Pansy, *V. tricolor*—both names suggestive of their leading characteristics. The Violets of S. America are shrubs.

The Order VIOLACEÆ, the Violetworts, includes the genus Viola, and 21 kindred genera, one of which, Solea, grows in the woods of New York, westward and southward.

Many of the Violets, especially those of S. America, possess valuable medicinal properties. Ionídium Itubu, called Poaya in Peru, is

---

* It is worthy of notice that when all the petals of a flower are equal in size and shape, they are also equally colored and streaked ; but as soon as one petal is enlarged for any special purpose, a change of color or adornment generally ensues. Thus in the Violets, especially in the Pansy, the lower petal forms an alighting place for the insect, and is more brightly colored than the rest of the flower—a door-step whence the color lines lead directly to the honey trove within, and in getting it the bee is sure to be dusted with the pollen.

† In ancient fable, Io, the daughter of Atlas, fleeing from Apollo, escaped to the woods, where, by the power of Diana, she was changed into a Violet, which even now modestly avoids the gaze of Phœbus by hiding her face in her own leaves.

one of the sources of Ipecac. Our native V. ovàta is a reputed remedy for the bite of the Rattlesnake. The common Sweet Violet of the conservatories (V. odoràta), the Ion of the Greeks, is famed for its fragrance. Its root is purgative, and employed in making the Syrup of Violets. The blue infusion of its flowers is employed by the French in numerous confections, and it also furnishes a chemical test, turning green with an alkali and red with an acid.

Scientific Terms.—Adnate anther. Cleistogene flower. Cordate. Crenate. Cucullate. Irregular flower. Innate anther. Lyrate Lyrate-pinnatifid. Reniform. Regular. Spur. Stipules. Auriculate.

XVII. CHICKWEED.

History.—We have before us a plant, humble in appearance, but of noble lineage and truly cosmopolitan. It is a common weed everywhere north of Mexico, and is abundant in Europe, whence it is supposed to have emigrated hither. It delights in cool, shady places on cultivated ground, and blossoms from the beginning of Spring to the end of Autumn.

Analysis.— THE LEAF REGION.—The root is annual and fibrous. The stem is slender and weak, and therefore prostrate or but half erect, nearly 1 foot long, with distinct *nodes* (joints), and terete *internodes* (p. 85), which are singularly distinguished by a hairy line which changes sides at each joint. Its branches are like forks—*dichótomous*. The leaves are ovate, smooth, entire, two at each node and opposite, 1' in length, mostly petiolate.*

The *Inflorescence* proceeds in the following order: 1st, the stem early terminates in a flower; 2d, a pair of branches arises from the axils of the upper pair of leaves

---

* It is curious to note how, as the chill of night comes on, the leaves fold together in pairs, enclosing the tender germ of the young shoot at the axil; while the upper pair but one is larger than the others and covers over the last pair, so securing the end of the branch.

and each terminates in a flower, leaving the first flower in the fork; 3d, the same process is repeated in each of these branches, the 2 secondary flowers being left below, each in a fork like the first, and so on to the last. Thus the central flower is the oldest, and the inflorescence is centrifugal.

The *Flower* may always be seen when the sun is shining, looking like a little star among the green leaves. It is regular, perfect, 5-parted, pedunculate. The green, hairy

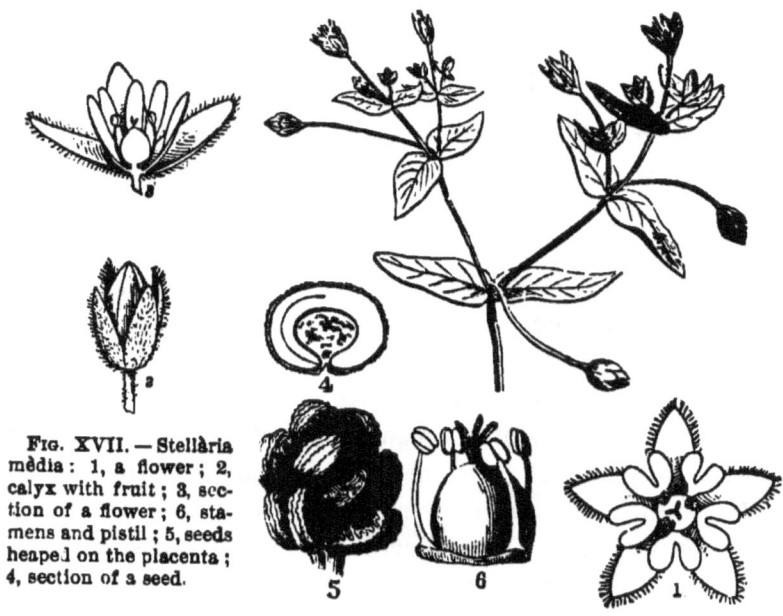

Fig. XVII. — Stellària mèdia: 1, a flower; 2, calyx with fruit; 3, section of a flower; 6, stamens and pistil; 5, seeds heaped on the placenta; 4, section of a seed.

calyx is larger than the white corolla. The sepals are lanceolate and quite distinct from one another. The petals are each deeply 2-cleft (*bifid*), so as to appear as if there were 10. The ovary is ovoid, surmounted by 3 sessile stigmas, and surrounded by the stamens, which are normally 10, each standing on a honey gland. But these little flowers assume large liberty; often, nay generally, their stamens are reduced to 5, 4, or even 3.*   Also late in the season they omit their petals, or develop some mere rudiments only.

The *Fruit* comes to be a capsule with 1 cell opening by 6 valves (or 3 split valves). The placenta stands in the center (*free-central*), covered with seeds which have a black, sculptured coat (*testa*), and a curved embryo around mealy albumen. (See Fig. XVII, 4.)

The Scientific Name of Chickweed is *Stellària mèdia*, the former in allusion to the silvery stars (*stella*, a star) of its blossoms, and the latter to its abundance (*media*, common). The chickadee and the chickens are fond of the plump seeds; hence the name *Chickweed*. The genus Stellaria is distinguished by having *5 bifid petals and 3 stigmas*.

XVIII. THE PINK.

Description.—The Garden Pinks and Carnations, so varied in form and coloring, are supposed to have descended from a single species known in Europe as Clove Pink, a native of the Southern Alps. In all its diversities it retains and is known by its glaucous evergreen foliage. We will take the common single Pink as the type.

Analysis.—The LEAF REGION is complicated, especially in the older plants. A *caudex* (a woody, leafless, close-jointed stem) with its root-end dissolving into fibers, divides above into many prostrate, tangled branches, which become herbaceous and leafy at their upturned ends. Here the true stems (*caulis*) begin, with lengthened internodes between the tumid nodes, bearing a pair of opposite, linear, sessile,

---

* The student will observe that the stamens come to maturity and shed their pollen before the stigmas are ripe. This prevents self-fertilization. (See p. 82.) The amount of honey secreted in these flowers must be infinitesimally small. By an elaborate calculation, Wilson concludes that it would require 2,500,000 florets like those of the White Clover to yield 1 pound of honey! This gives some idea of the industry of the bee, and the amount of labor represented in every honeycomb.

STELLARIA. 81

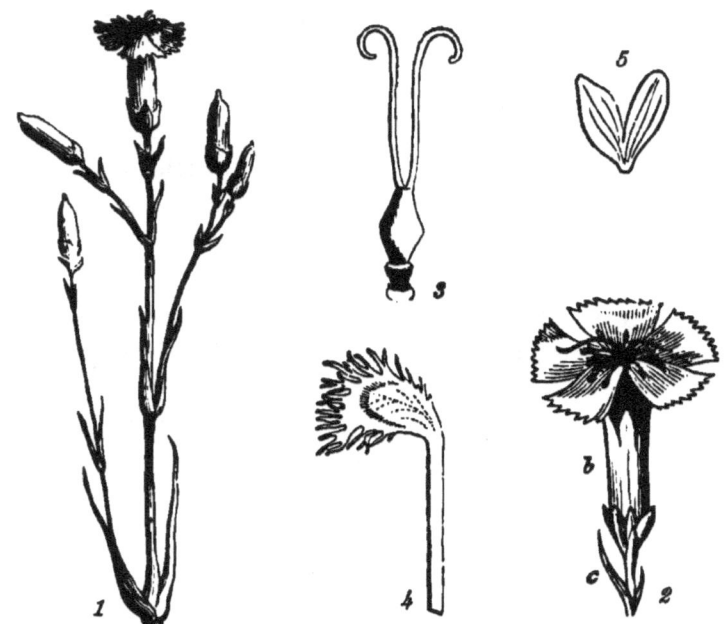

Fig. XVIII.—Diánthus Caryophyllus: 2, a flower, showing all the organs; 3, the ovary and two styles; 4, a petal of Silène stellàta, *fimbriate* (fringed) and *unguiculate* (petiolate); 5, a petal of Ceràstium, *bifid*.

apparently veinless leaves at each joint, and a terminal (centrifugal) inflorescence.

THE FLOWER REGION.—The flowers, few and large, on account of their peculiar grace and elegance, have been celebrated in story and song from the earliest times. The green calyx, of 5 united sepals, as seen by their disunited ends, is truly a *flower-cup* or *vase*, supported at the base by 2 or 3 ovate bractlets. The 5 petals, arranged with consummate art, insert their long claws (petioles?) within the vase, forming a fringed and tinted corolla. Likewise the 10 long stamens and 2 styles. The ovary is but one, becoming a 1-celled, 4-valved, capsular fruit. The many black seeds with embryo but little curved, and mealy albumen, cover the *free-central* placenta.

THE PINK.

FERTILIZATION.—The nectar, situated in the deep narrow calyx, can be tasted only by the long tongue of Moths and Butterflies. The stamens usually appear first, issuing from the throat of the blossom, and after showering their pollen on the heads of the visiting Moths, wither away; immediately, the 2 long recurved styles emerge, ready to receive the pollen brought from the other flowers.*

TERATOLOGY.—Carnations are Pinks made double by artificial culture. A careful analysis reveals the curious change which has taken place. The petals are multiplied to about 20. The stamens have divided themselves each into several, all more or less deformed, but evidently likewise tending toward the shape of a petal. The ovary may have become triple, with a third style, and the calyx may have burst open. This unruly behavior is called *teratology* (*teras*, a monster). See Lesson XXVIII.

The Name is *Diánthus Caryophýllus*. Dianthus (*Dios, anthos*) means the Flower of Jove, or God's own flower; and Caryophyllus, the Clove Tree, is applied to this species on account of its peculiar fragrance. The genus Dianthus is known by a *tubular, bracted calyx, and two styles*.

Classification.—The two genera—Dianthus and Stellaria, represent the Order of the Pinkworts, or CARYOPHYLLACEÆ. The student will remember that they coincide in the following characters:

    Herbs with swollen joints and opposite, entire leaves.
    Flowers regular, symmetrical, 4 or 5-parted.
    Petals distinct, or wanting.
    Stamens twice as many as the sepals, or fewer.
    Ovary compound, free from the calyx.
    Embryo curved or coiled on mealy albumen.

---

* Plants with this habit of promoting cross-fertilization will be found quite numerous, and are called *proterandrous* (from the Gr. *proteros*, earlier, *andres*, stamens). On the other hand, other plants mature their pistils earlier than their stamens. The Plantain, for example, pushes out its long hairy style a day or so before its own stamens are ready, in order to receive pollen from other flowers.

The **Pinkworts** thus defined will include 35 genera and 1000 species, growing on mountains, rocks, hedges, and waste places, in the temperate and cold regions of the World. Except for ornament they seem to be of little service to Man.

Sweet William (*Diánthus barbátus*), with flowers in dense cymes, and infinite variety of color, is from Europe.

Catchfly (*Silène*) is noted for the *viscid rings* just below the joints, serving not only to catch little flies and gnats, but to stop the ascent of ants who would steal the nectar intended for the bees.

Corn Cockle (*Lychnis Githágo*) is a handsome weed growing in Wheat fields because its seed cannot be winnowed from the grain.

Soapwort (*Sapondria*), called also Bouncing Bet, flourishes by roadsides. It has large handsome flowers, and its herbage when bruised may be used for soap.

Scientific Terms in Lessons XVII and XVIII: Bifid. Caudex. Caulis. Ditchotomous. Free central placenta. Internodes. Nodes. Proterandrous. Teratology.

XIX. THE WILD GERANIUM.

Description.—In May and June, the forests are everywhere adorned with the large, round, pale-purple flowers of the Wild Geranium or Cranesbill. Beautiful in itself, it is invested with additional interest by its associations.* It stands firmly erect, 1 or 2 ft. high, clothed with whitish hairs.

Analysis.—The *Root*. Under the soil we find a stout fleshy root-stock or rhizome, with many strong fibers attached, a very astringent taste, and evidently perennial. It is often sought by the country people as a household medicine, and sold in drug-stores, by the name of Cranesbill.

The *Stem* arises in Spring, terete, *jointed*, and with a few leaves on long radical petioles. At each joint (*nodus*)

---

* There is an Eastern tale that the Geranium was formerly a Mallow, but Mahomet having laid a garment upon it to dry, it was transformed into this more beautiful plant. A marvelous change indeed; for the two plants are botanically unlike by many grades of difference.

THE WILD GERANIUM.

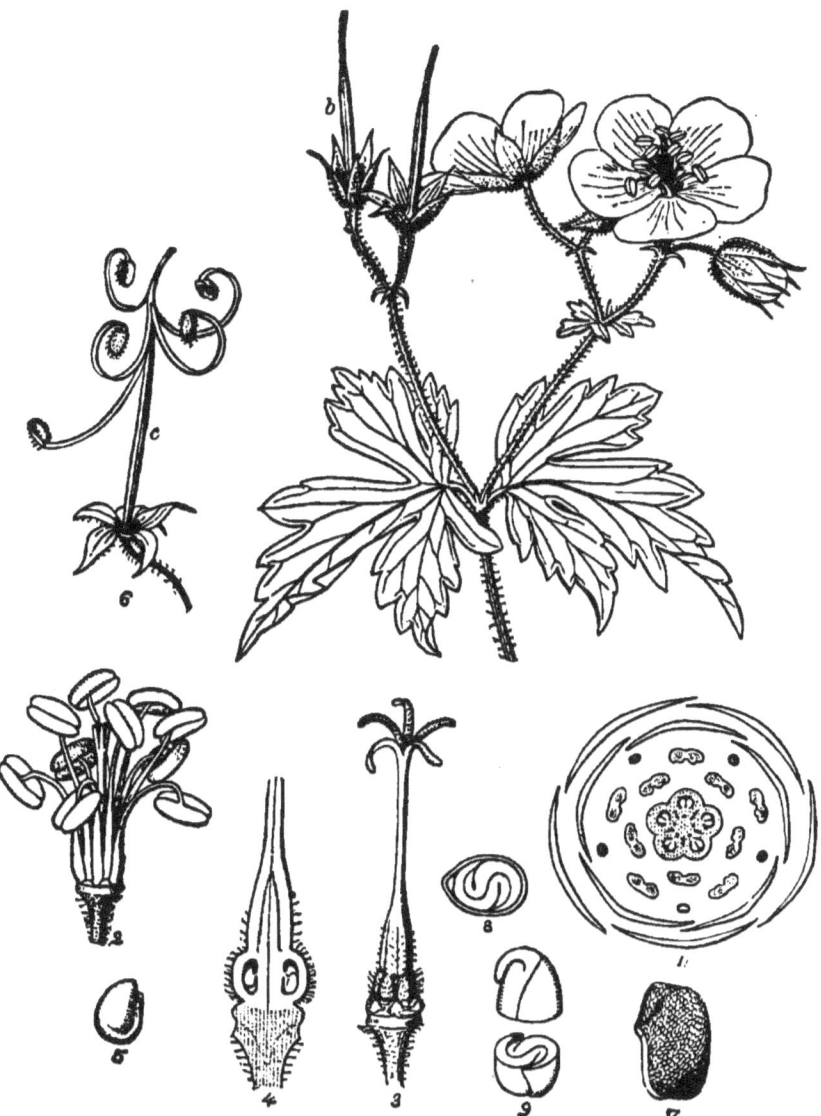

FIG. XIX.—Geranium maculàtum: 1, plan of the flower; 2, the stamens, and (immature) pistil; 3, the mature pistil; 4, section of ovary; 5, ovule; 6, fruit; 7, a seed; 8, 9, embryo.

the stem bears a pair of opposite leaves and divides or *forks* into branches. Botanists call the joints *nodes*, and the portion of stem intervening between the nodes, *internodes*. In Geranium the nodes are conspicuously swollen. In most other plants they are slightly swollen and bear but one leaf.

The *Leaves* are palmi-veined, and palmately 5 or 7-lobed, the lobes cuneate below and cleft above. Each petiole is furnished with a pair of narrow, acuminate stipules at the base.

The *Flowers*, regular and symmetrical, are an inch broad and 5-parted throughout. The green sepals are 3-veined, and awn-pointed; the petals obovate, bearded at the base on the short claw (*unguis*); the stamens ten (2), alternately a longer and a shorter one, with the anthers *versatile*, i. e., balanced on the tip of the filament; the pistils 5 cohering into one (3).* The torus is remarkable. It bears 5 *glands* alternating with the petals, and supports a central column rising in the midst of the styles to their top. It is the *carpophore*, or fruit-bearer (6, *c*).

The *Fruit* (6) is a *regma* (fracture), so named from its curious behavior. The entire compound pistil persists, grows into a slender column (*b*) having the 5 ovaries at the base. When fully ripe, it breaks up into its 5 constituent carpels, and each carpel is then borne upward on its recurving elastic style, which still remains attached to the top of the carpophore. In this position it is inverted, and its black dotted seed (7) drops out.

---

* It has often been observed that the stamens of this plant mature sooner than their pistil. When the flower first opens, the style is short and the 5 stigmas close up as seen in Cut 2. After the anthers have shed their pollen, then the stigmas arise and spread out ready, but too late to receive it. Now they must get their supply from other and later blossoms. Such flowers are called *proterandrous* (Note, p. 82). Cross-fertilization is evidently the end of this arrangement.

The *Seed* has a rough shell (*testa*) entirely filled by the embryo whose 2 cotyledons are nicely folded together and bent over on the radicle (8, 9). There is no albumen.*

THE PLAN of the flower (1) shows 6 circles, each with its 5 members all alternating: 1st, the sepals; 2d, the petals; 3d, the honey glands; 4th and 5th, the stamens; 6th, the ovaries.

The Name *Gerànium* comes from the Gr. *géranos*, a crane, because of a fancied resemblance of the fruit to the beak of that bird. The species in hand is *G. maculàtum*, or Spotted Cranesbill, named for the pale blotches often seen on its leaves. Another common species is *G. Robertiànum*, the Herb Robert, with smaller and redder flowers.† These and 100 other similar species *have perfectly regular flowers, with ten perfect stamens, and the fruit a regma.*

XX. THE HORSESHOE GERANIUM.

Description.—Let us now interrogate that popular house-plant, the Horseshoe Geranium (known by the brown ring on its rounded leaves), and learn whether it be indeed a Geranium, or of some other genus of this splendid Order.

Analysis.—THE LEAF REGION.—The plant before us was reared from a cutting; hence its roots are artificial and give no proof of their native form. The stem lives and grows from year to year, becoming a woody branching shrub with a greenish bark.

It is said that in seedlings the earlier and lower leaves are

---

* In seeds where the albumen is wanting, the seed-lobes or cotyledons become thick and fleshy with starchy matter, infolding the embryo for its protection while sleeping, feeding it with their own substance in its early growth, and finally appearing, as usual, a pair of leaves, the *first* which the plantlet unfolds.

† The pretty flowers are roseate and penciled with purple. The leaves are more finely divided and cut, emitting a strong odor when handled. Late in the season they are subject to a parasitic fungus, appearing sprinkled with darkish specks.

opposite. In our plant, however, all are alternate, with long petioles and broad stipules. The blade is *orbicular* or nearly round, palmi-veined, with many shallow lobes, green, but liable to endless markings and shadings.

THE FLOWER REGION.—The peduncles issue opposite to the leaves and grow much longer than they, bearing an umbel of 12 or more flowers, with an involucre of 6 bracts. The flowers are an inch broad, 5-parted, and slightly irregular. Of the 5 green sepals, the upper one protracts its base down the pedicel, forming a slender tube upon it, or a slender spur adhering to it (*s*). Of the 5 scarlet petals, the 2 upper are somewhat smaller than the 3 lower.- Of the 10 filaments, only 7 bear efficient anthers. The pistil and fruit are nearly as in the Wild Geranium save the twisted beaks.

FIG. XX.—Flower of Pelargònium zonàle.

The Name.—Now, with its irregular, spurred flowers, its 7 perfect stamens, can this plant be a true Geranium? The French botanist, *L'Heritier*, A. D. 1787, separated such plants, and formed a new genus with the analogous name, *Pelargònium* (Storksbill, Gr. *pelargos*). It now includes 170 species, all native in S. Africa, and many favorites alike in the conservatory and in the humble cottage window.* Ours is *P. zonàle.*

The Record of the analysis of Wild Geranium is to be used as a monitor, not a guide. The form of the tablet is like those in the Plant Record. The letters following the

---

* Another group of Gerania having *regular flowers with only 5 good stamens and the awns of the carpels twisted and barbed*, was separated from the Linnæan genus by L'Heritier and named Erodium (Heronsbill). One of its species, *E. cicutarium*, deserves mention as a forage plant of great value. It is rare in the Atlantic States, but in California overspreads hill and plain to an immense extent. It is called *Al-fìlirèa*. It starts from seed annually, grows rapidly, feeds flocks and herds during Winter and Spring on its sweet herbage, and, in the dry Summer and Autumn, on its nutritious seeds left broadcast on the ground.

name of the organ are the initials of the categories heading the page; e. g., "Root, L. K." stands for Root, its Life and Kind; "Leaves, L.P.C.F.S.Q." is for Leaf or Leaves, their Life, Place, Construction, Form, Size, and Qualities. Pelargonium may be recorded in like manner.

Scientific Terms.— Awn-pointed. Carpophore. Claw. Cleft. Glands. Internode. Node. Orbicular. Regma.

ORGAN, (its)	Life, Habit, Number, Place, Dehiscence, Kind, Construction, Form, Placentation, Size, Qualities, Appendages.
Plant, L.H.S.Q.	♃, Herb erect, 1—2 ft., with whitish hairs.
Root, L.K.	♃, Root-stock thick, with many fibers.
Stem, L.H.K.F.	⊙, erect, brachiate, caulis-jointed, terete.
Leaves, L.P.C.F.S.Q.	⊙, opposite, petiolate, stipulate, palmate, 5—7-lobed.
Inflorescence, P.K.A.	Terminal, cymous, centrifugal, involucre 2-leaved.
Flower, N.C.	5-parted, perfect, complete, regular.
Calyx, F.Q.	Bell-form, green, ciliate.
Sepals, L.N.P.F.	Persistent, 5, imbricate, bristle-tipped, oblong.
Corolla, F.Q.	Rosaceous, lilac-purple.
Petals, L.N.P.F.	Deciduous, 5, contorted, rounded, claw bearded.
Stamens, N.P.C.	10, hypogynous, filaments slender.
Anther, D.C.F.	Longitudinal, 2-celled, innate, oblong.
Style, N.C.F.	5 united, slender, around the carpophore.
Stigma, N.F.	5, linear, stellate.
Ovary, C.F.Pn.	5 united carpels, separating in fruit.
Fruit, N.D.K.F.Q.	5, breaking up, a regma, beaks curved.
Seed, N.C.F.Q.A.	1 in each carpel, oval, black.

LOCALITY.—*Woods.* (Date), *May 3, 1877.*
CLASSIFICATION,—**PHENOGAMIA**, POLYPETALOUS EXOGENS.
—Order, GERANIACEÆ.
NAME.—Latin, **Geranium maculatum.**
—English, Spotted Cranesbill.

XXI. THE YELLOW WOOD SORREL.

Description.—The Yellow Wood Sorrel, with clover-like leaves, is almost ubiquitous. It blossoms from May to September, in open places, from the Great Lakes to the Gulf, and even to the Pacific Ocean. There are other plants

Fig. XXI.—Oxalis stricta : 1, plan of the flower; 2, vertical section of flower.

of its kind more beautiful, but none more instructive nor so generally within the reach of the botanical student.

Analysis.—THE LEAF REGION. From perennial creepers under ground, or from the seed, a slender stem arises, simple and erect at first, but soon branching and reclining.

90   YELLOW WOOD SORREL.

The leaves are arranged *alternate* on the stem, compound, *trifoliolate*, consisting of a long petiole, with 3 *leaflets*. The form of these leaflets is a study. They are broad and notched above so as to present 2 rounded lobes at the apex —the cordate form inverted (*c*), or *obcordate*. The venation is also to be studied, whether the leaf be *palmate-trifoliolate*, with the leaflets all sessile alike (as if cut from a palmi-veined leaf), or *pinnate-trifoliolate*, with the terminal leaflet

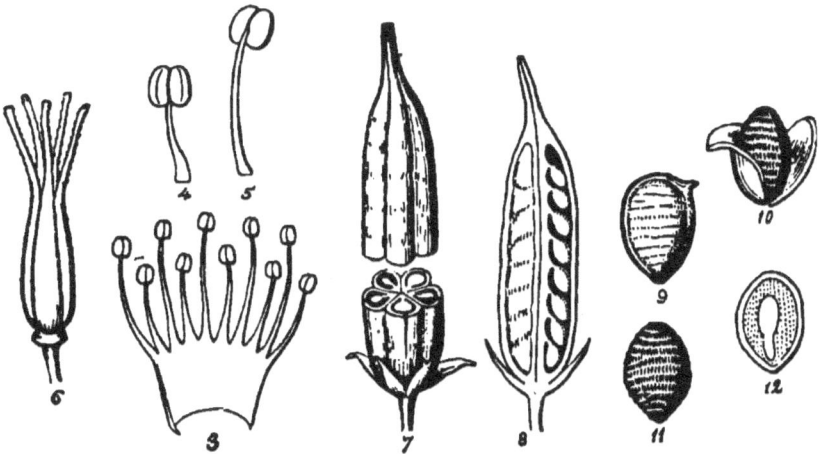

Oxalis stricta: 3, the united stamens; 4, 5, stamens; 6, the 5 pistils; 7, 8, sections of the fruit; 9, seed, 10, testa, 11, naked; 12, embryo.

stalked as in Bulbous Crowfoot (p. 50), or better, in the garden Bean. This question let the reader decide.

Inflorescence.—Next arises the inquiry, Where is the *axil* of the leaf? It is analogous to the arm-pit—the inner angle between the petiole and the stem (*a*). From this point issues the bud which develops into a branch, and in this case, at least, the peduncle which bears the umbel of flowers. Hence the inflorescence is *axillary* and *umbellate*.*

THE FLOWER REGION.—The flowers of Sorrel, like those

---

* Not truly umbellate, however, as it becomes a *cyme* when the flowers are more than 3.

of Geranium, are 5-parted throughout; sepals 5, petals 5, stamens 2 × 5, and pistils 5, with their styles distinct and their ovaries united (6). The stamens are also united at the base, or *monadelphous* (*monos*, one, *adelphos*, brotherhood), (3).

Æstivation.— Here recurs an interesting topic suggested by the curious posture of the petals when just opening. How are they folded together in the bud? The inquiry is important, since many genera and even some orders are characterized by their mode of æstivation. (See the plan, 1.) In reference to the Wood Sorrel the student would instantly reply, "The petals are *twisted* in the bud!" And the terms *contorted* or *convolute* express the idea. The petals are all rolled together in one direction, each having one edge within and one without. The same is true of the flowers of Cranebills, and generally of Storkbills.

Sleep.—At the approach of night, or in cloudy weather, these flowers close up and fold their contorted petals again as in the bud. So also do the leaves. At night each leaflet falls back on the stalk, folds its two halves together face to face, and thus remains asleep, as it were, until awakened by the morning sun.*

---

* The *vigils* of plants are evidently dependent on the degree of light; but different species are variously affected. While many, like Oxalis, open and close with the day for many days in succession, others, like the Morning Glory or the Portulacas, open for a day only, and perish. The evening Primrose opens at 6 o'clock P. M. for a night, perishing at sunrise; and the Four-o-clock at about that hour for a few hours only, and the Water Lily (Nymphæa) opens and reopens only while the sun is high, from 8 to 2. And there are other plants which, like the Gerania, open once for all, and close not by night or day, until they close forever.

Recent researches show that the sleep of plants generally bears some relation to their peculiar wants. Thus the Daisy closes its flowers and hangs its head when night comes on, or the gathering clouds forbode a storm, lest the dew or rain dissolve the nectar stored up in its florets. The Nipplewort (Lapsana), common in Europe, opens before six and closes before ten in the morning, in order that the bees who are early risers may taste its nectar, and not the ants, who delay until the dew is off, and would not scatter its pollen. Again, night-flowers are adapted to the habits of certain nocturnal moths which are needful agents in their fertilization. Such flowers are always white or pale yellow, the only colors visible in the darkness.

The *Fruit* is an oblong capsule (7) made up of 5 carpels, each with a row of seeds in its cell (8). The carpels open on the back (*dorsal dehiscence*) and do not separate from the central axis (*carpophore*) at once as they do in the regma of Geranium. The seed is anatropous (9), with a loose, separable outer coat (10, 11) and a large straight embryo buried in albumen (12).

THE PLAN (1) shows the sepals to be *quincuncial* (p. 43) and the petals *contorted* in æstivation.

The Name of this plant, *Oxalis* (*oxus*, sour), refers to the taste of the herbage given to it by the presence of oxalic acid in the form of a salt (binoxalate of potash). The specific name, *O. stricta*, alludes to its upright stem; the other species being mostly acaulescent. Oxalis is an admirable genus, embracing in all lands 220 species, many of which are beautiful conservatory and house plants.*

Classification.—The student can hardly fail to notice the striking resemblance of the Oxalides to the Gerania. Their flowers are completely analogous. The fruit in both consists of 5 carpels—as many as the sepals, attached to a central axis arising from the torus. Oxalis takes rank, therefore, with the Gerania in the Order Geraniaceæ.

Scientific Terms.—Alternate. Axil. Axillary. Contorted. Convolute. Dorsal dehiscence. Leaflets. Monadelphous. Obcordate. Palmate-trifoliolate. Pinnate-trifoliolate. Trifoliolate. Umbellate.

---

* One of the most popular is O. floribunda (Lehmann) from Brazil. A specimen growing in our study has bloomed five months continuously, displaying some 300 roseate flowers on every sunny day. It is very exacting in its vigils, closing its leaves at sunset, and its flowers always except in the sunshine.

XXII. THE JEWEL WEED.

Description.—There is a tall, smooth herb, with pellucid, jointed stalks, abundant in low swamps and along shaded rivulets, variously called Jewel-weed, Snap-weed, Touch-me-not, &c. Fresh specimens, together with the cuts, will show how much a flower may differ from its kindred and still be recognized.

Analysis.—THE LEAF REGION.—The annual root; the juicy stem, with its tumid nodes; the ovate, serrate leaves; and the axillary inflorescence, present no new features. The student unaided may readily characterize them. But the flowers and fruit are remarkable.

THE FLOWER REGION.—The *Flowers*, although so very irregularly and oddy developed, are evidently in nature and intention 5-parted. Their color is a deep orange spotted with reddish-brown. Only 4 sepals appear, but the upper one (See, 2, *s*) is notched at the broad apex, showing it to be double, or composed of 2 sepals united. The lower (*y*) is a conical hood (*cucullate*), or a cornucopia, tipped with an *inflected* spur. Only 2 petals appear (*p p*), but each has a lobe and is evidently composed of 2 united petals. There are 5 short stamens with *introrse* (p. 40) anthers, and bearing 5 scales covering the stigma, which is sessile on the ovary. A vertical section (1) gives an inside view of the flower.

The *Fruit* is a general wonder. In form and structure (3) it resembles that of Oxalis; in behavior, it is very different. At maturity its 5 muscular carpels or valves become elastic springs ready to break loose at the slightest touch. Coiling with a sudden jerk they fly from the central axis (4) and scatter the seeds in every direction.*

---

* This is one of the many devices for the dispersion of seeds in which intelligence and wisdom are manifest. The seeds of Maple and Ash are furnished with wings for

94   THE JEWEL WEED.

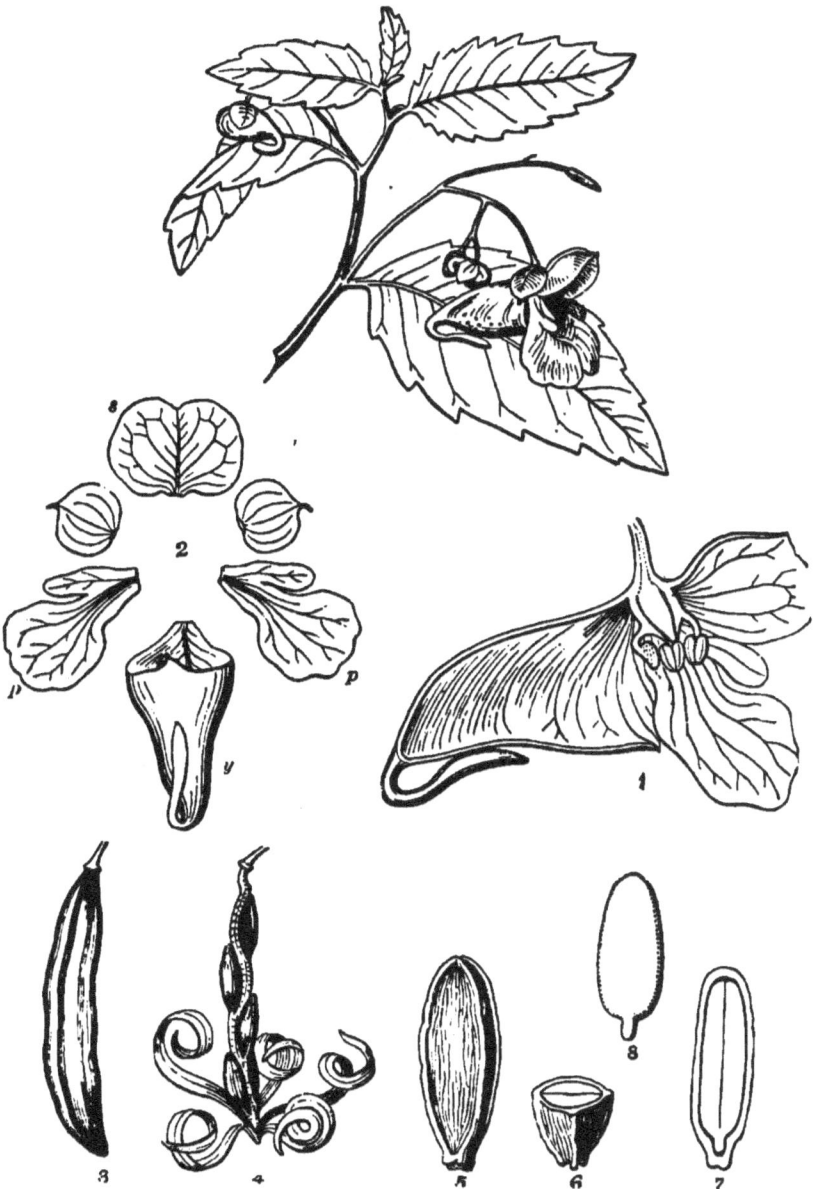

Fig XXII.—Impàtiens fulva: 1, a flower dissected; 2, a flower displayed; 3, ripe fruit; 4, the same just exploded; 5, a seed; 6, section; 7, 8, cotyledons of the embryo exposed.

The *Seed* appears in (5) to be anatropous; in (6) and (7), dicotyledonous, without albumen, the large straight embryo filling the testa. (8) is the naked embryo.

The Name.—In this connection, let the Garden Balsam be analyzed. Though very different in general aspect, we shall find that the above description of the Jewel-weed applies to it in almost everything but color and clothing. Both are species of the genus *Impátiens* (impatient). The Jewel-weed is *I. fulva* (Nuttall), the Balsamine, *I. Balsamína* (Linn.).*

Classification. — How do these plants resemble the Gerania?—In their *tumid nodes, 5-parted flowers, 5-carpelled fruit, elastic carpels, central axis, and in the spurred lower sepal*, here *free* from the pedicel while in Storkbills adhering to it. These are marks of the Order Geraniaceæ.

XXIII. THE NASTURTION, OR INDIAN CRESS.

Description. — This old and popular garden flower assumes a style of beauty intermediate between the Gerania and the Jewel-weed. It is a native of Peru, whence it was brought nearly 200 years ago. Its study will reveal several new forms of structure, both in leaf and flower. It is an annual herb, or with protection, *biennial;* but the root per-

---

this purpose. Tick-seeds and Burr-seeds are provided with hooks and barbs by which they lay hold of men and animals and are thus, by unwilling agents, scattered far and wide. The seeds of Thistle, Dandelion, Silkgrass, made buoyant by means of their downy appendages, are wafted afar, beyond rivers, lakes and seas. The Squirting Cucumber, as it ripens, becomes distended with water until at last it breaks from its stem and projects through the rupture, with amazing force, the mingled seeds and water. Rivers and Ocean currents are always transporting seeds from country to country. Thus the Cocoa and the Cashew nut and the seeds of Mahogany have been known to perform long voyages without injury to their vitality. Squirrels laying up their winter stores in the earth, birds migrating from clime to clime and from island to island, conspire to effect the same important end.

* Only 2 species are native in N. America, 1 in Europe, 1 in Siberia, 1 in Madagascar, and 100 in India. All are remarkable for the elastic bursting valves of their pods.

ishes with the stem. It is cultivated from seed both for ornament and use.

ORGAN.	Life, Habit, Number, Place, Dehiscence, Kind, Construction, Form, Placentation, Size, Qualities, Appendages.
Plant, L.H.S.Q.	☉ *herb terrestrial, erect, tall (3—5 ft.), smooth.*
Root, L.K.	*Annual, axial.*
Stem, L.H.K.F.	*Herbaceous, erect, branching, terete, with tumid nodes.*
Leaves, L.P.C.F.S.Q.	*Deciduous, alternate, pinni-veined, petiolate, simple, ovate, serrate, 1—3' long, smooth.*
Inflorescence, P.K.A.	*Axillary, racemed, pedunculate, bracted.*
Flower, N.K.C.S.	*5-parted, perfect, irregular, unsymmetrical, 1' long.*
Calyx, F.Q.	*Saccate and spurred, colored like the corolla.* [*inflected.*
Sepals, L.N.P.F.	*Deciduous, 4, imbr., upper double, lower a cornucopia, spur*
Corolla, F.Q.	*Much suppressed and deformed, orange-color, spotted.*
Petals, L.N.P.F.	*Decid., 2, imbric., spreading, double, unequally 2-lobed.*
Stamens, N.P.C.	*5, hypogynous, short, each bearing a scale.*
Anther, P.D.C.F.	*Innate, introrse, dehisc. lengthwise, ovate.*
Style, N.C.F.	*Very short or none.* [*filaments.*
Stigma, N.F.	*5 united into 1, sessile, 5-lobed, covered by the 5 scales of the*
Ovary, C.F.Pn.	*Compound, superior, 5-celled, oblong.*
Fruit, N.D.K.F.Q.	*A 5-valved capsule, oblong, opening elastically.*
Seed, N.C.F.Q.A.	*Several, exalbuminous, with straight embryo.*

LOCALITY.—*In wet woods, Hoboken, N. J.* (Date), *May 27, 1877.*
CLASSIFICATION.—**PHENOGAMIA.**
  ORDER.—GERANIACEÆ, THE CRANEBILLS.
    NAME.—Latin, **Impatiens fulva.**
      —English, Touch-me-not. Jewel-weed.

Analysis.— The *Stem* is slender and weak, trailing along the ground, or climbing, not by twining, but by the help of its leaf-stalks. Thus it may arise 3 feet, or protected from frosts in the house, 6 feet.

The *Leaves* have the form of a shield or target (*pelta*) called *peltate*. The roundish or angular blade is attached to its stalk not by its margin, but by a point within. It is a singular form; but if you compare it with a leaf of the Horseshoe Geranium (p. 87) you will doubtless conclude that it results from the cohesion of the 2 base lobes. The same thing occurs in the Ivy Geranium. The long petiole, when its help is needed in climbing, coils about the supporting object like a tendril, as in that plant also.

The *Flower*. All parts of the 5-parted irregular flower are alike colored, orange or variegated. The upper sepal is united at the base with the other 4 and produced backward into a spur. The petals are inserted in or on the throat of the spur, the 2 upper sessile, the 3 lower fringed (*fimbriate*) at the base and supported on a claw (*unguis*), or *unguiculate*. There are 8 unequal stamens, and 3 ovaries around the central axis or style. (See Fig. XXIII, Appendix.)

The *Fruit*. The ripe fruit contains 3 large, fleshy, ribbed, 1-seeded nuts, such as we often see upon the table as a substitute for Capers.*

Classification.—This plant is sometimes called Trophywort, its leaves and flowers being likened to shields and helmets. For a like reason the generic name is *Tropæolum* (*tropæum*, a trophy). The species is *T. major;* i. e., the Greater Trophywort. *Its flowers 5-parted and spurred, its stamens unsymmetrical, and its 1-seeded, separable carpels,* ally it to the Storkbills and the Order Geraniaceæ.

The Order Geraniaceæ, as now constituted, associates 16 genera and nearly 750 species. But the association is not truly natural, and

---

* The true *capers* are the flower-buds of Cápparis spinòsa, a shrub of S. Europe, preserved in vinegar.

the genera are often too discordant for a happy family. They therefore resolve themselves into several clans or suborders. The five genera last treated, viz., Geranium, Pelargonium, Oxalis, Impatiens, and Tropæolum, represent at least four of these suborders, which for a long time were regarded as Orders. (See *Botanist and Florist*, p. 67, flora.) The following formula, brief and easily remembered, will, with few exceptions, characterize all the Geraniaceæ:

Herbs or shrubs.
Flowers perfect, symmetrical.
Stamens as many or twice as many as the sepals, often some of them abortive.
Carpels and cells as many as the sepals, separating from a persistent axis or carpophore.
Seeds few, with no albumen (except in Oxalis).

Scientific Terms.—Fimbriate. Peltate. Suborder. Unguiculate.

XXIV. THE SHEPHERD'S PURSE.

Description.—This is a homely little weed intruding itself into gardens and fields everywhere unbidden, yet illustrating the principles of Botany and the mysteries of vegetable life quite as well as loftier plants. Beginning to blossom in early Spring, it continues developing flower after flower as it rises higher and higher, until fruit and flower together embellish the long racemes.

Analysis.—THE LEAF REGION.—The student will now require no further aid in recording the analysis of the root and stem; the longevity, venation and inflorescence; the presence or absence of stipules, petioles, bracts, hairs, and branches; and the position and arrangement of the leaves.

The *Leaves* are of two forms. The radical are oblong and *pinnatifid* or feather-cleft; the cauline are *sagittate* or arrow-shaped (*sagitta*, an arrow), and *amplexicaul* (stem-clasping). Here observe, whence do the branches arise? (p. 90). What is the position of the racemes? What is the procession of the flowering?

FIG. XXIV.—Capsélla Bursa-pastòris: 1, the flower; 2, the stamens and pistil; 3, the pistil alone; 4, the pistil seen edgewise; 5, the silicle; 6, the same open, showing the seeds; 7, a seed; 8, 9, embryo, with cotyledons incumbent.

THE FLOWER REGION.—The *Flowers*, as a whole, (1) are 4-parted, regular, and *unsymmetrical* (for the special organs are not all of this *radical* number 4, nor multiples of it), viz., sepals 4, petals 4, stamens 6 (2), pistils 2 (3, 4), united and stigma double, with a short, thick style. Observe the relative length of the stamens (2); 2 of them are shorter than the other 4—a fact denoted by the term *tetradynamous*. Also the special form of such corollas or flowers is *cruciform* (*crux*, a cross), and resembles, when the petals are spread, the Maltese cross.

The *Fruit*. As the raceme is the oldest at its base, there we must look for the earliest fruits. Their curious shape reminds one of a leathern pouch—the shepherd's purse, of course. Their form is obcordate. Their slender pedicels are longer than when in flower. A thin narrow partition within divides them crosswise into 2 cells, and at length they break into as many boat-shaped or *carinate* valves, liberating the seeds. Such a fruit, when short as in this case, is called a *silicle* (a little pod); when long as in Mustard, a *silique*.

The *Seed* in its testa appears as in (7), with its embryo bent double; also in the cross-section (8), and the naked embryo (9). But the radicle is so bent as to lie over on the *back* of one of the cotyledons, not on its edge. So this seed is said to be with *radicle incumbent*. It has no albumen.

The Name, *Capsella* (a little box or capsule), is applied to the genus. The specific term, *C. Bursa-pastòris*, is the same meaning in Latin as in English.*

Scientific Terms.—Amplexicaul. Carinate. Cruciform. Pinnatifid. Radical number. Sagittate. Silicle. Silíque. Tetradynamous. Unsymmetrical.

---

* Indeed the name as well as the plant seems to be truly cosmopolite. The traveler who sees little else to remind him of his native soil, can generally find the homely Shepherd's Purse growing by the wayside. It is abundant even amid the classic ruins of Rome, and there too the peasant calls it "Borsa de Pastor."

XXV. THE TOOTHROOT CRESS.

Description.—This plant frequents the rich woodlands of the Northern and Western States, by the streams and fountains, blooming in the spring months. It is *glabrous* (smooth), 1 foot high, and often called Pepper-root.

Analysis.—The *Root-stock*, by its peculiar shape, suggests the former name, and by its crisp, pungent taste, the latter. It is long, creeping, white and fleshy, with many knobby, tooth-like projections, and is sometimes broken up into a string of knobby tubers.

The *Stem* with its two opposite trifoliolate leaves and terminal raceme, presents no new features.

The *Flowers* are constructed on the plan described under Capsella, but are large and showy. When the 4 oblong white petals are expanded, their mimicry of the Maltese cross justifies the term cruciform so generally applied to this class of flowers.

The *Fruit* affords a new field of study. But if your specimens are not well matured, search for riper ones. The form is outwardly a contrast with the silicles of Capsella. The pod is many times longer than wide, lanceolate-linear, broadest near the middle and tapering below to the pedicel and above to the style. It is composed of 2 carpels and opens by 2 valves. Within, it is divided lengthwise into 2 cells. Each cell nourishes a row of seeds attached alternately to opposite sides of the valve or partition.

Notwithstanding the difference in external form, this fruit agrees in structure with the silicle of Capsella, and its name, *silique*, is of similar import: *Silicle* being a short pod and silique a long one.

The *Seed*, skillfully dissected, as seen in (4, 5), shows the embryo bent double, so that the radicle rests on the edge of

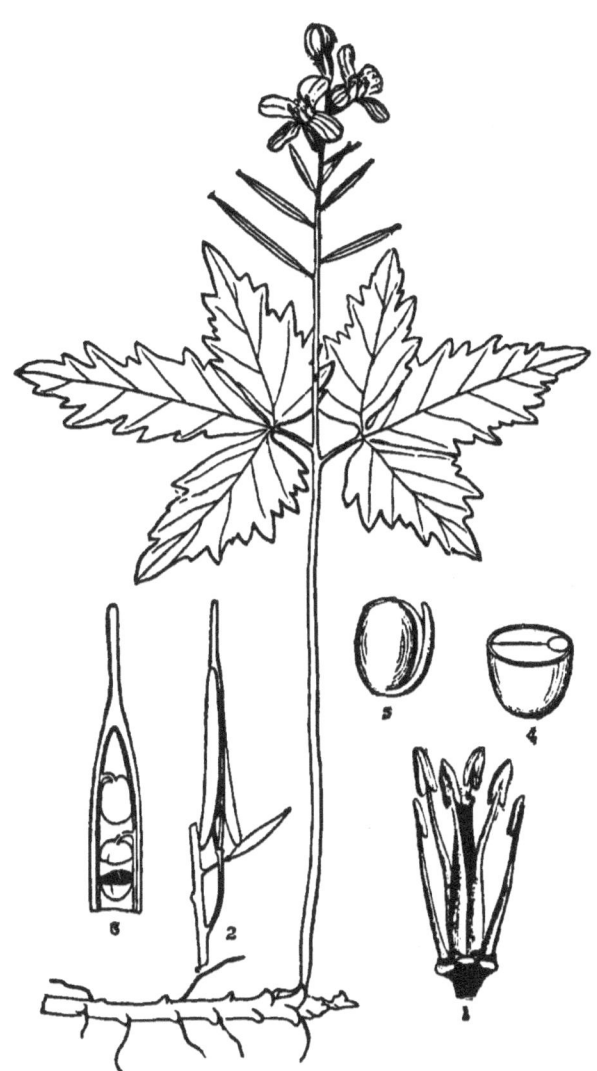

Fig. XXV.—Cardamine diphylla: 1, the stamens, and 1 style; 2, a silique; 3, seeds in the silique; 4, seed cut across; 5, the embryo—cotyledons accumbent.

CARDAMINE. 103

the two cotyledons. The phrase *cotyledons accumbent* is applied to seeds so constructed. Compare this with the seed of Capsella.

The Name of this plant is *Cardamine diphylla*, or the Two-leaved Toothroot. There are other species having 3 whorled leaves, and still others with alternate leaves.*

Classification.—Here let the student take note of the affinities of these two genera, Capsélla and Cardamìne, in the following points. Thus will he learn the characters of a large and important Order, the CRUCIFERÆ, or Crucifers.†

>Herbs, without stipules.
>Inflorescence centripetal, bractless racemes.
>Flowers cruciform, perfect.
>Stamens didynamous, hypogynous.
>Ovary double.
>Fruit siliques, long or short.
>Seeds exalbuminous, with the embryo bent double.

The Order of Crucifers is truly natural, and embraces about 172 genera and 1600 species, chiefly in the Temperate Zone. More than 100 species are peculiar to this Continent. Among them are nutritious vegetables, as Cabbage, Turnip, Radish. Some are condiments in general use; as Horse-radish, Mustard. The bland *Rape-seed oil* is expressed from the seeds of the Rape (*Brássica Napus*). Woad, a blue dye, is obtained from the root of *Isàtis tinctòria*.‡ In medicine the Crucifers are stimulant and antiscorbutic, but none are poisonous. They all contain a volatile acrid principle abounding in sulphur and nitrogen, which is the cause of the unpleasant odor they emit in decaying. Here too belongs many a favorite garden flower, like Sweet Alyssum, Candytuft, Wall-flower (*Cheiranthus*), Honesty (*Lunaria*), and Stock (*Matthiola*).

---

* The Toothroots were first named by Linnæus, *Dentària* (*dens*, a tooth). The original species were easily distinguished from the genus Cardamine. But other species recently found in California combine the characters of both genera, so as to unite them into one, taking the *older* name. Some authors, however, still retain the genus Dentària, and call our plant D. diphýlla. This is therefore its *synonym*.

† In this connection, let the Mustard plant, Wall-flower, Pepper-grass, Candytuft, &c., be analyzed and registered.

‡ This dye is famous in history as having been employed by the Britons in staining their bodies in order to frighten their enemies.

Scientific Terms.—Cotyledons accumbent. Cotyledons incumbent. Glabrous. Silicle. Silique. Synonym.

XXVI. THE STRAWBERRY.

Description.—May, charming May is the festival of the Roseworts. Now trees and shrubs, as well as tender herbs, are bursting into bloom, adorning field and forest. So many and varied are the flowers asking attention, that one is bewildered in choosing. Among the Roseworts let us first examine the Strawberry plant.

Analysis.—The *Root* and *Stem* are, as in Liverleaf and Blue Violet, subterranean. But the stem (crown, p. 53) at certain times sends out from its top a slender, terete, red *runner*, one or more, a foot in length, tipped with a bud, which on touching the soil, develops roots downward and leaves upward, and so founds a new plant.

The *Leaves* are complete in their organization, having blade, petiole, and stipules—the blade palmate-trifoliolate as in Oxalis (p. 90). The leaflets are ovate, oval or obovate, coarsely *serrate*, having teeth pointing forward like sawteeth, and, like the scapes, pubescent, with soft appressed hairs. The petioles are *villous*, with coarser spreading hairs.*

The *Scape* branches irregularly into a cluster, of which the central flower is the oldest; hence the inflorescence is *centrifugal*, progressing from the center outward, and the cluster, a *cyme*.

---

* The hairs of plants constitute an interesting study. They are composed of a single long cell, or of a transparent tissue of cells placed end to end like a string of beads. There is an endless variety in their length, abundance, and quality. Sometimes they are soft and close like down, sometimes stiff and rough like bristles. Now they form a fringe like an eyelash, and now they silver the surface with a silky gloss. Here they curve backward into a hook, oftentimes barbed. In the Nettle, they are hollow stings with a bag of poison concealed. In the Sun-dew, they are tipped with a glistening exudation like a dew-drop. They warmly clothe the early catkins of the Willow, and decorate the landscape in the waving plumes of the Pampas Grass. Cotton, a great staple of commerce, is but the hair with which a seed is fledged.

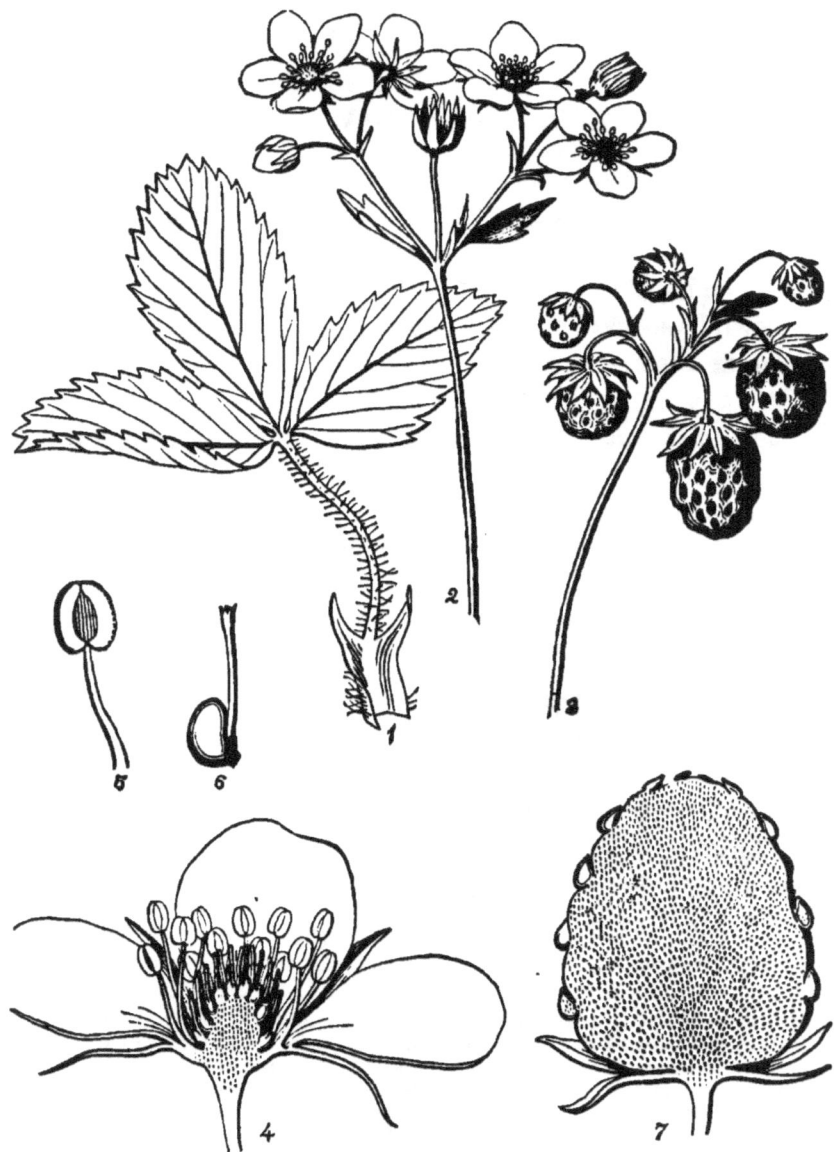

FIG. XXVI.—Fragària vesca : 1, a leaf with its stipules ; 2, a cyme ; 3, fruit ; 4, vertical section of a flower ; 5, a stamen, innate ; 6, a pistil with lateral style ; 7, vertical section of a strawberry.

The *Flower*, in its general plan, resembles the Buttercup; but have you not already taken note of two remarkable differences? The 5 green sepals are here reinforced by 5 similar alternating bracts, appearing like a double calyx or a calyx of 10 sepals. The ∞ (= many) stamens, a multiple of 5 (at least in the wild plant), are, in situation, *perigynous* (*peri*, around, *gynè*, pistil), i. e., adhering at the base to the calyx as if inserted on it (4). How does this compare with the flower of Buttercup? It is an important distinction. The ∞ pistils, situated as in Buttercup, are peculiar in form (6), with a *lateral* style, and quite distinct from one another.

The *Fruit* is a strawberry; it needs no other name, for there is no other like it. It consists of the enlarged pulpy torus (7) bearing on its surface the many one-seeded carpels —the achenia, the true fruit of the botanist.* While in bloom, the flowers are erect and above the leaves, but in fruit they nod and ripen in partial concealment.

Æstivation. The 5 white petals, like those of the Buttercup, are *quincuncial*, i. e., 2 are wholly outside, 2 are wholly within, and 1 oblique, or half without and half within. Compare this with the flower of Oxalis.†

The Name, *Fragària*, alludes to the fragrance of the luxurious fruit. Two species, *F. Virginiàna*, and *F. vesca*, grow wild in woods and fields. Under cultivation, the pulpy torus is wonderfully enlarged.

Scientific Terms.—Complete leaves. Cyme. Imbricate. Perigynous. Pubescent. Quincuncial. Runners. Serrate. Villous.

---

* In the vegetable economy the pulpy deposit in fruits has reference to the dispersion of the seeds rather than their nourishment in germination. It feeds and nourishes the birds, which in turn plant afar off the seeds which they have swallowed, while man avails himself of only its superabundance. It is interesting to note the varieties of form and place which this deposit takes in different fruits. In the strawberry, the delicious morsel is in the torus; in the raspberry it is in the achenia; in the blackberry, in both torus and achenia. In the checkerberry, the calyx contains the rich deposit; in the grape, the pericarp, and in the apple, both calyx and pericarp, while in the pineapple the whole inflorescence becomes gorged with pulp.

† The term *imbricate* is more general, applying to both these special forms in which the petals overlap each other like shingles. (See p. 43, Note.)

XXVII. THE APPLE TREE.

The Tree.—The transition from the humble herb to the lofty tree is sufficiently abrupt; but except in growth and stature, the real difference may be slight. While the herb devotes its entire annual income to its offspring, the tree reserves a portion for itself, treasuring up solid wood in its stem and branches.

The *Trunk* is the appropriate name for the stem of a tree—one of the most interesting and useful of all natural objects. In the Apple Tree, it is short and definite, seldom more than 7 or 8 feet high. At the base in the ground, and at its summit, it suddenly terminates, dissolving into roots strong and far-reaching below,* and into branches, branchlets and spray above, forming the rounded, aerial *head*. This kind of trunk is termed *solvent*, in distinction from the *excurrent* trunk, as shown in the Pines (p. 216).

The *Wood,* seen in cross-sections easily made with a saw and plane (8), displays, 1st, the pith in or near the center; 2d, the purple *heart-wood*† around it; 3d, the white *sap-wood* around the heart-wood; 4th, the *bark* around all; 5th, the *annual layers* or wood-rings, here two only, of which the outer is the younger; and 6th, the silvery *medullary rays* running from the pith (*medulla*) to the bark. Each layer is the growth of a year; consequently the number of the layers suggests the age of the branch, and a similar section of

---

* If all the roots of growing plants could be laid bare of earth, the sight would be marvelous. It is roughly estimated that an Elm is as large below as above ground. What shall we say of the root of the common Red Clover, which has been known to descend a distance of five feet; or a stalk of Wheat which, within forty-seven days after planting, sent down its fibers into a light subsoil seven feet? The roots, blindly searching around after food, often seem to be endowed with some special sense.

† More properly called *duramen* (*durus*, hard). It is heart-wood only in respect to situation, for it bears no part in the life and vegetation of the plant. It is more the seat of death than of life; hence it often decays, leaving the trunk hollow while the tree is as flourishing as ever. Thus the tree at once both lives and dies, like the Coral, which is dead below and alive at the extremities.

108   THE APPLE TREE.

the trunk indicates the age of the tree. They also show that the wood grows externally, for the new layer is deposited outside the old wood next to the bark. In other words, the mode of its growth is *exogenous* (*exo*, outside, *genao*, I grow). Compare the growth of the Palm (p. 225).

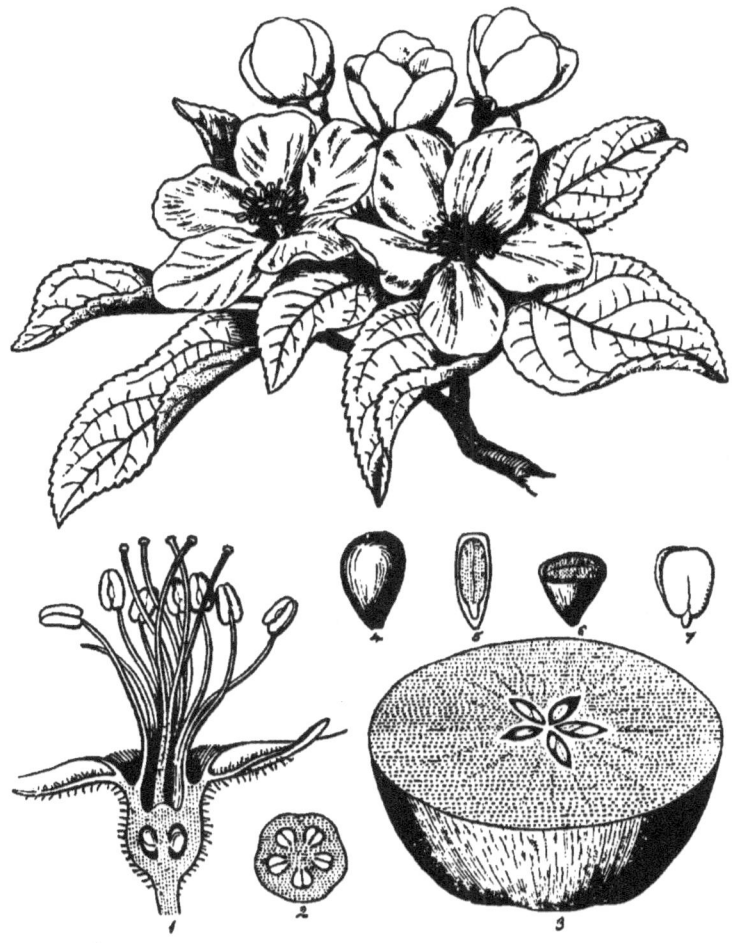

Fig. XXVII.—Flowering branch of Pyrus Malus: 1, section of a flower; 2, section of the ovary; 3, section of the fruit (apple); 4, a seed; 5, 6, sections of same; 7, the embryo.

PYRUS.   109

THE FOOD OF PLANTS.—Whence and what are the materials for sustaining this growth? Learn from the treatment which your house plants receive. Their roots are immersed in a pot of soil. You shower them at night with water containing a little added ammonia. You open the

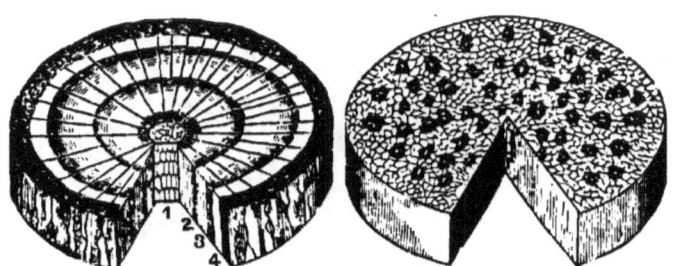

8, cross-section of an exogenous stem of 2 years' growth: 1, pith; 2, 3, annual layers of wood; 4, the bark and white new layer (*cambium*) under it. 9, an endogenous stem (Indian Corn), with no layers nor bark.

windows in the morning to bathe them in fresh air. Then, with warmth and sunshine, they ask no more. So the tree, by its myriad of roots and rootlets, imbibes water containing ammonia and various mineral matters in solution. Thence this *sap*, as we call it, creeping from cell to cell of the root, stem and branch, and dissolving the sugar, gum, &c., it finds on the way, finally reaches the leaves. Here is the chemical laboratory of the plant.* Much of the water having performed its work of carrying up the raw material from the earth, evaporates through the pores of the leaf.† Through

---

* It is curious to notice how the trunk and branches of the tree are all the work of the frail and transient leaf. Slowly, year after year, generation after generation, it is steadily elaborating, from air and rain and sunshine, these solid structures which are to remain its enduring monument, when it has faded and crumbled to dust.

† It has been found by experiment that the leaves of plants exhale moisture to an enormous amount. An acre of beets, during a single day of sunshine, evaporates from 17 to 19 thousand pounds of water. A Chestnut tree 35 years old, in 24 hours, lost over 63 quarts of water. The upward pressure of the ascending sap is very great. Experiments were made, in 1720, by Dr. Hales of England, proving that this force in a Grapevine was equal to the weight of a column of water 43 feet high. Similar experiments were made in 1873, by President Clark, of the Massachusetts Agricultural College, on a native vine (Vitis æstivàlis). On May day, a mercurial gauge was

these same pores the leaf inhales the air, and now under the influence of the sun (see *Chemistry*, pp. 97, 181, 237) the sap is converted into a thin mucilage which contains all the elements of vegetable growth. The sap then descends * and spreads through the tree, especially along the inner surface of the bark, supplying every want of the young layer of wood, of the leaf and the flower.

Analysis.—The *Leaves* of the Apple Tree are complete, having a pair of subulate (awl-shaped) stipules at the base of the short petiole. The blade is ovate, serrate, and beneath *tomentous* with a dense covering of matted hairs. Its venation is pinni-veined and reticulated.

The *Inflorescence* is an umbel issuing from one bud, with no peduncle and therefore sessile.

The *Flowers* are pedicellate, regular, 5-parted, polyandrous, perigynous, rose-white, fragrant; the 5 sepals are so united below as to form an urn-shaped fleshy tube which adheres to and encloses the 5-carpelled ovary (1, 2); † the 5 petals are broadly oval, quincuncial (p. 43), inserted by their short claws with the ∞ perigynous stamens (p. 106). The 5 styles are partly united.

The *Fruit* is a *pome* (3). Mark how it is crowned with the persistent calyx lobes (sepals), a proof that the pome consists of the enlarged calyx-tube with the enclosed ovary, both gorged with pulp. Make a cross-section (2, 3) and see the 5 cells with cartilaginous walls, and the circular greenish line around them in the pulp marking the boundary between

---

attached to the severed end of one of its main roots. At first there was a suction downward, gradually diminishing until the 10th. Thence until the 29th, an upward pressure increased and attained a force equal to the weight of 88 feet of water!

* We can easily prove the existence of this descending current, for on making an incision into the bark of a young branch, the sap will ooze from the upper and not the lower lip of the cut.

† Thus the ovary is apparently situated *below* the calyx, whence it is said incorrectly to be *inferior*, and the calyx *superior*. The phrase *ovary adherent*, or *calyx adherent* are of the same meaning and more correct. In all the flowers heretofore analyzed the *calyx is free* (*inferior*) and the *ovary free* (*superior*).

the ovary and the calyx tube. In each of the 5 cells are 2 seeds, each large enough for an easy analysis (4). The brown *testa* outside is readily separated from the soft, white inner coat (*tegmen*). The cut (7) shows the naked embryo, with its radicle and two cotyledons; (5) and (6) are sections.

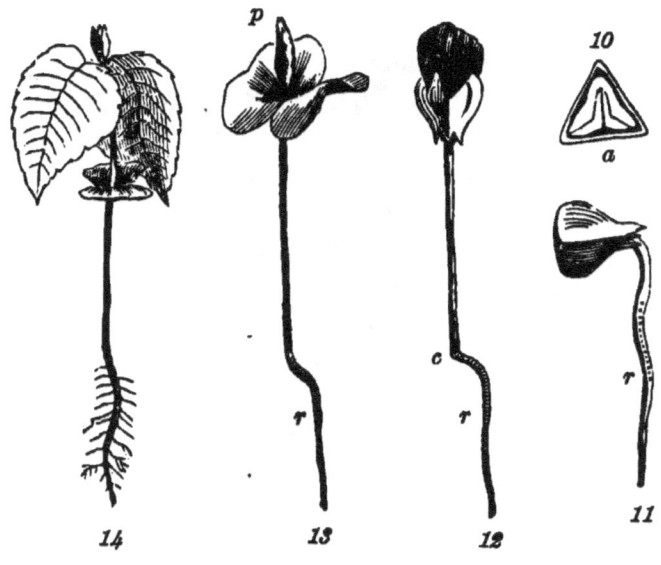

10, germination of the Beech-nut: 10, cross-section showing the 2 folded cotyledons; 11, the radicle only protruded; 12, the ascending axis above *c* appears; 13, the cotyledons expand into a pair of leaves—the first two, and show the plumule; 14, with rootlets and the first leaves from the plumule.

Germination. Plant these seeds, or find in the garden seeds already sprouted (*germinating*), and learn what the several parts become in the plant. Here is seen the radicle, *r*, growing downward as the root, the *plumule*, *p*, growing upward as the stem, and the 2 cotyledons appearing and acting as the first leaves. The store of food laid up in them is serving (like the albumen where there is any) for the nourishment of the plantlet until with roots and leaves of its own it becomes able to provide for itself.*

The Name, *Pyrus,* the Latin word for Pear, was adopted by Linnæus as the title of a genus including the Pear, Apple, and other trees. The specific name, *Malus,* is the ancient Roman term for Apple.

Scientific Terms.—Adherent calyx. Adherent ovary. Annual layers. Bark. Cambium. Exogenous. Excurrent trunk. Free calyx. Free ovary. Germination. Heart-wood. Inferior calyx. Inferior ovary. Medullary rays. Pith. Plumule. Sap-wood. Solvent trunk. Superior calyx. Superior ovary. Tegmen. Testa.

XXVIII. THE ROSE.

Description.—Among flowers the Rose reigns supreme. Without it no garden, however humble, is thought complete. For its dignity, fragrance, and infinitude of form and color, it is interwoven with all poetry and art.† The species growing wild in the whole world may be 120, while the garden varieties are numbered by thousands. Double Roses are the delight of the florist, and very instructive; but they are unfit for regular analysis. You must bring the Wild Rose of the swamp or prairie, or the Sweet Brier of the field.

---

* Here the analysis of the Pear, Peach, and Cherry flowers will be in order, also the Yellow Cinquefoil (Potentilla). Compare the flowers by making vertical sections, and you will find striking analogies as well as contrasts. In Raspberry, the torus and its ovaries are elevated above the calyx; in Cinquefoil, they are on a level; in Rosa (1), depressed far below it.

† The Rose was a great favorite with the Greeks and Romans. Nero caused showers of Roses to be sprinkled on his guests at banquets, and Heliogabalus carried this to such an extent that several persons were suffocated before they could extricate themselves from the mass. The Sybarites, it is said, slept on couches stuffed with Rose petals. This flower was dedicated to the god of silence, and a Rose hanging over a guest-table was a hint that conversation was to be "sub-rosa." It was customary for wreaths of Roses to be worn by warriors, while Rose-leaves (petals) were strewn on the dishes on festal occasions, and the bushes were planted on graves as a mark of respect and love. In later times the Rose was especially dedicated to the Virgin, and in Dante's Paradise she is termed the "Mystic Rose,"

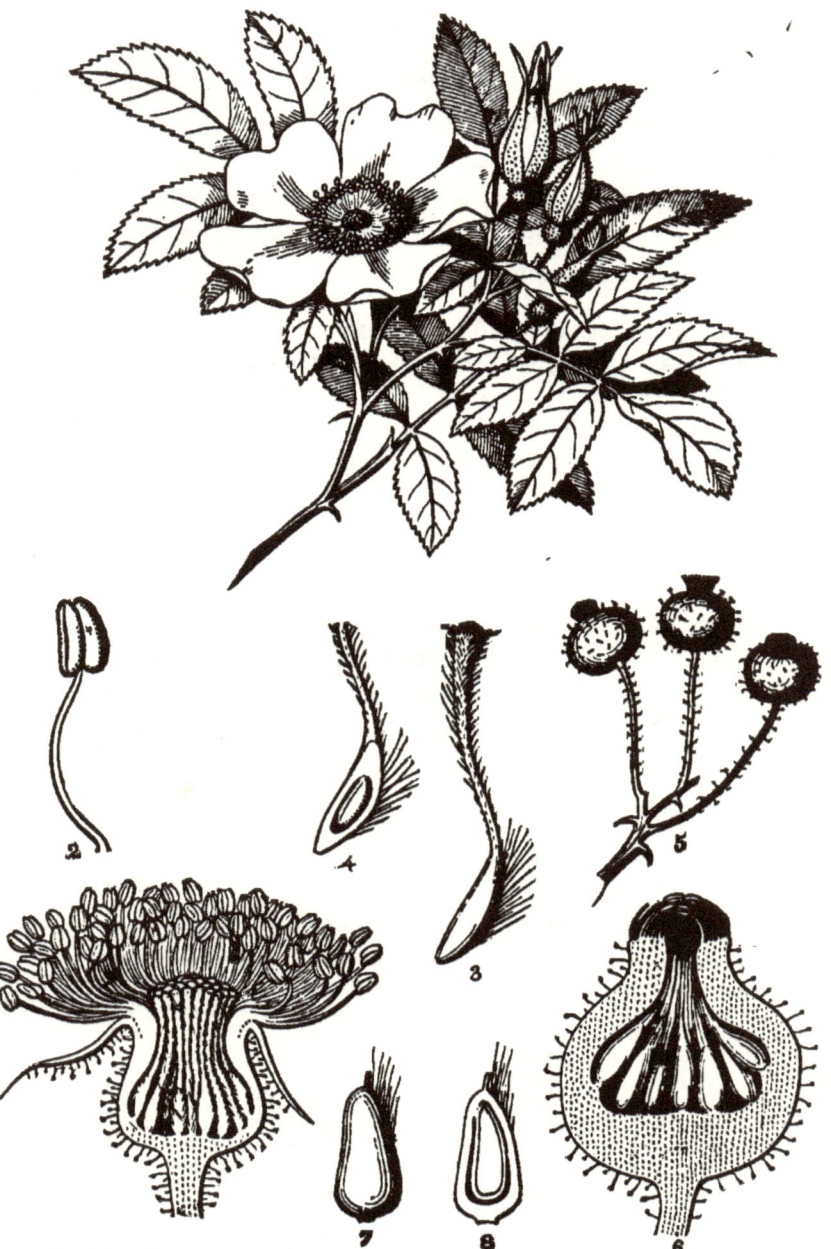

Fig. XXVIII.—Rosa Carolina : 1, vertical section of flower (minus the petals) ; 2, a stamen ; 3, a pistil ; 4, showing the 1 ovule ; 5, the hip (fruit) ; 6, section ; 7, an achenium ; 8, showing the suspended seed,

Analysis.—We have in hand a *shrub*, with woody stems, 4 to 6 feet high—your own stature more or less. It is beset with *prickles*. Mark their structure; compare them with the *thorns* of the thorn-apple, and note how they differ.

The *Leaves* are *odd-pinnate*, consisting of 5 to 9 (an odd number) elliptical, serrate leaflets, with pinnate and reticulate veins. The 2 narrow stipules are adnate to the petiole.

The *Inflorescence* is in the form of a *corymb*—the peduncle branching into unequal pedicels bearing the flowers at about the same level.

The *Flower* may be cut vertically (1) for a better view of the structure. The calyx tube is seen *inclosing*, but not adhering to the 15—20 distinct, *inferior* (?) ovaries. The 5 sepals are some or all of them tipped with a leaflet. The 5 quincuncial, *broad-obcordate, rose-colored* petals are inserted with the ∞ perigynous stamens on the calyx tube.

The *Fruit*, generally called a *hip* (5, 6), is globular, fleshy, red, inclosing (not adhering to) the 15 or more distinct bony achenia. The styles are persistent (3, 4). The seed is anatropous, *suspended* from the upper part of the cell (7, 8).

THE DOUBLE ROSE.—In wild native plants double flowering is extremely rare. Have you ever known an instance? This phenomenon seems to be an unnatural condition induced and perpetuated by the art of the cultivator. Its study reveals many secrets of nature's laws. In any collection of Roses we find some single and natural, some semi-double, and some fully double. The 1st exhibits a corolla of 5 broad, equal petals, and a great number of stamens within its enclosure. The 2d displays a 5-petaled corolla with 20 or more narrow petals superadded, and in their midst a diminished number of stamens. The 3d shows a hundred petals filling the entire space within the original corolla. What has become of the stamens? Look again at the semidouble Rose. You find the stamens in a state of transition, as it were; some perfect, yellow; some with a slight red expansion on one side, others on both sides; some again half stamen and half petal, and in all degrees of progress—plainly indicating whither the stamens are going and have gone. From this study, the nature of the double Rose, and the tendency of the stamens become manifest. In the semidouble, a part of the stamens have been transformed to petals, and in the double, all of them.

The metamorphosis often goes still further. In that curious variety, the Green Rose, the stamens have all reverted first to petals, and then to leaves. In the Damask Rose, we have occasionally seen a leafy branch occupying the place of the stamens and pistils. Similar changes are continually occurring not only in Rose, but in Pæony, Camellia, Balsamine, Violet, and other plants, and all agree in teaching that the stamen is a leaf modified and adapted to a special purpose. The student will look for further illustration of this interesting doctrine, which was first suggested by Linnæus about A. D. 1750.

The Name *Rosa* is of Latin origin. The wild species just described and portrayed is *R. Carolina*.

Classification.—The Strawberry, Apple Tree, and Rose, as we now see, are allied to one another and to the Order of the ROSACEÆ, or Roseworts, by the following characters:

Stipules present.   Stamens ∞ perigynous.
Flowers regular.   Seed anatropous.
Corolla quincuncial.   Embryo straight.
Albumen none.

The Roseworts, moreover, having the *embryo 2-lobed, their flowers 4 and 5 parted, their leaves net-veined, and their wood, if any, growing* *by annual external layers,* are classed, with the Crowfoots, Crucifers, Cranebills, &c., in the province of the EXOGENS.* It is a large and important Order, including 71 genera and 1000 species, arranged in several suborders (see *Botanist and Florist,* p. 101). They are chiefly natives of the N. Temperate regions. Their prevailing property in bark and root is astringency. Prussic acid occurs in the Almond and Apple suborders. Many of the species produce edible fruits.

The Peach tree (*Amýgdalus pérsica*) is a native of Persia. The Nectarine is a variety of the same species. In recent botanies it is *Prùnus vulgàris.* The Wild Plum of our own forests is *Prùnus Americàna.* The Garden Plum (*P. doméstica*) is a native of Europe. The Cherries are also various species of *Prùnus.* The Cherry Laurel (*P. Caroliniàna*), a beautiful evergreen tree of the S. States, has so much prussic acid in its leaves and cherries as to render them poisonous. The seed of the Peach is poison for the same reason.

The Blackberry (*Rùbus villòsus*) is powerfully astringent. *R. strigòsus* is the delicious Raspberry ; *R. occidentàlis,* the Thimble-berry; *R. odorátus,* the Mulberry. *R. spectábilis,* the Shadberry, bears the finest fruit in Oregon.

The Attar of Roses, an essential oil of exceeding fragrance, is distilled from *Ròsa Damascèna* and *R. moschàta.* 20,000 flowers are

---

* It will now be seen that from the leaf alone, or from the smallest fragment of it, the place of a plant in the natural system of classification can be determined. It is the venation of the leaf that affords the criterion, and this pervades the fragment as well as the whole. We have now considered three diverse modes or types, which are severally characteristic of the three Grand Divisions of the Vegetable Kingdom. *First,* the forked-venation of the Cryptogams, best seen in the Ferns (p. 21). Here the veinlets divide and subdivide each into 2 smaller ones, which run on straight from center to circumference—terminating in the margin or in a fruit-cluster, never reuniting when once parted. This is the simplest of all kinds of venation, and is peculiar to the simplest of all plants which rise above the purely cellular Mosses, where there are no veins at all. *Second,* the parallel-venation of the Flowering Endogens, seen in the Tulip, and the Grasses. Here the veins run parallel to each other on the surface, without dividing or interlacing, so that the leaf may be torn from base to apex regularly along the course of any of the veins. Such an arrangement of veins, comparatively simple, is associated with flowers always ternate in their parts, seeds always with one cotyledon in its embryo, and a stem without bark or annual woody layers (p. 33). *Thirdly,* the netted-venation of the Flowering Exogens, just studied in the Buttercups, and now seen in the Apple-tree. In such leaves the venation becomes intricate. The veins divide to infinity and their ramifications reunite as often, forming a network all through the leafy tissue, as beautifully illustrated in "skeleton leaves." This, the highest type of venation, is associated with the highest development of vegetable life—flowers many-parted, seeds with two cotyledons, and wood (if any) with bark and annual layers.

required to make a rupee's weight (one-half ounce), which sells for $50.

To the genus *Pyrus* belong the Pear Tree, Apple Tree, and Medlar. *P. coronaria*, with flowers as fine as the Rose, is our Wild Crab Tree.

The Quince (*Cydònia vulgàris*) is a native of Austria. *C. japónica*, the beautiful Japan Quince, is from Japan.

The *Spireas* are always conspicuous in the gardens and parks, as well as the Roses and Japan Quince. So also the Hawthorns and Mountain Ash.

Scientific Terms.—Corymb. Double flower. Metamorphosis. Obcordate. Prickles. Shrub. Suspended ovule. Thorns.

XXIX. THE PEA AND ITS TRIBE.

Description.— Of this large and important Order we have no plant more characteristic than the common Garden Pea. It is also represented by the Sweet Pea, Wild Pea, Locust, and Wistaria, which may be studied in this connection.

Analysis.—The pupil will answer queries like the following: What is its *term* or period of life? What is its habit?* How does it climb? What is the composition of its leaves? What the outline of the leaflets?—of the stipules? Has it *stipels* (little stipules at the base of each leaflet)?

The Tendrils. A new feature now appears. Tendrils are growing from the extremity of the rachis of the pinnate leaves, and they are themselves compound. Each tendril consists of 3 or more coiling threads or fibers—aids to the plant in climbing.† Leaves thus furnished are called *cirrhous*.

The *Inflorescence* consists of peduncles springing from the axils, each bearing 2 or more white flowers.

---

* Habit denotes the form, appearance, and conduct of a plant, as it would strike the general observer, without reference to scientific accuracy.

† The action of a tendril looks almost like intelligence. It remains extended, and straight, with only a slight curve or hook at the extremity, as if blindly searching for some object to lean upon. If such support is not soon found, it often sweeps around

The *Flower* is nodding, 5-parted, and irregular after a fashion termed *papilionaceous* (butterfly-shaped). There are 5 sepals, united at the base and free above. Of the 5 petals (1) the upper and odd one is the largest, and in the bud covers all the rest. It is called the banner or *vexillum*. The others are in pairs; the 2 lowest being the keel petals (*carinæ*); the 2 intermediate, the wings (*alæ*).* Of the 10 stamens (2), 9 are united by their filaments, while the 10th is separate and free, a condition termed *diadelphous*, that is, in 2 sets. The pistil is one only, and that (3) a simple carpel with one style and stigma. The style is bent at a right angle, and flattened as if laterally, with a groove on the back and a bearded line in front next the free stamen.†

7, a legume.

The *Fruit* is a *legume*. It is a dry pod, oblique in form, one-celled, 2-valved, opening at both its edges (*sutures*) and having the seeds in one row along the front suture—not along both as in a silique (p. 101). Much is learned by a careful study of the legume. Open it at the front suture. The two valves still conjoined at the back will represent a leaf, with seeds like buds developed

---

horizontally, describing in an hour or two a complete circuit, like the free end of the twining stem or branch of the Morning Glory (p. 186). When at length the tendril, with the hook at its extremity, touches a twig or other object, it immediately twines about it, while at the same time in its middle portions it coils up on itself, as if by shortening its own length to draw the plant closer to its support. The mechanical difficulty of coiling up while fixed at both ends is overcome by its turning in opposite directions, causing a sharp angle where its course is reversed. This is best seen in the Gourd Tribe and the Grape-vine. After its hold is thus secured it grows strong and tough; but if it fails to reach its object it soon droops, coils up, and perishes.

* In the Pea tribe the banner is brightly colored so as to justify the name, since when expanded it cannot fail to attract insects. The lower petals are frequently joined together in one piece, forming a kind of doorstep on which the insects may readily alight. On their attempting to enter the throat of the flower, by springing open the alæ, the stamens beneath are liberated and are dashed with some force against the insect's body, so as to cover it with pollen.

† In the Sweet Pea, cf the genus Lathyrus, the style is flattened on the back and in front; and this circumstance alone separates these two genera.

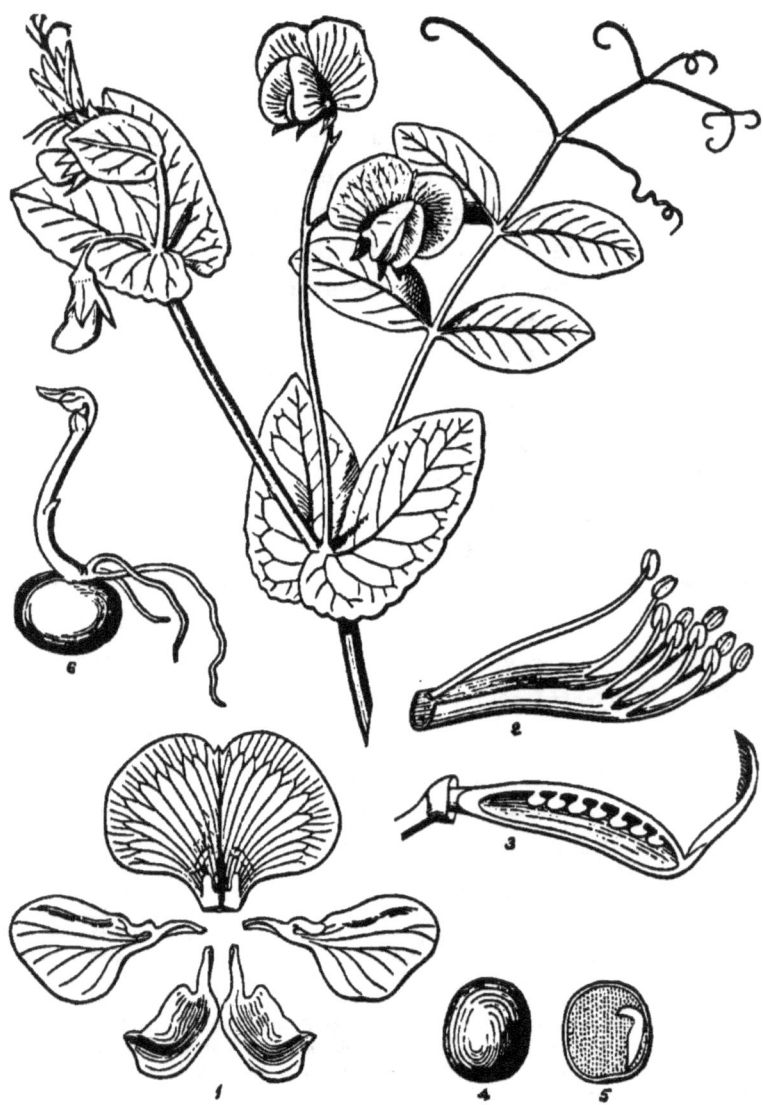

Fig. XXIX.—Pisum sativum: 1, the corolla displayed; 2, the diadelphous stamens; 3, the ovary dissected, and the peculiar style and stigma; 4, a seed; 5, the embryo with one of the cotyledons; 6, a seed germinating.

along the margins, as if a leaf were transformed into the pistil, and (like the leaf of a Begonia when planted) produced buds at its edges.*

The *Seed* is so large (4) that it may be analyzed without a microscope. Remove the testa and you find within, the embryo alone, as in the seed of the Apple. This consists of 2 large cotyledons, and between them (5) a conspicuous *radicle* and *plumule*.

GERMINATION.— Out of the thick cotyledons, the radicle and plumule draw their first nourishment and wake into life and growth. In the figure (6), the radicle has sent forth 3 rootlets tending downward, and the plumule, a stem tipped with a bud tending upward.

The Name is the ancient Latin *Pisum; P. sativum*, that of the species, indicates that the seed is sown in gardens. Its native country is unknown.

Scientific Terms.—Alæ. Banner. Carinæ. Cirrhous leaves. Diadelphous. Front suture. Keel petals. Legume. Papilionaceous. Rachis. Stipels. Vexillum. Wing petals.

---

* The fruit, as well as each organ of the flower, is a modified leaf, or leaves. The simple fruit, formed of a single pistil, like the achenium of the Crowfoots or the legume of the Pea, is a single leaf. It is folded upward so that its upper surface becomes the inner, and its united edges the placentæ where the seeds are developed. In the Peach, another simple fruit, the upper skin of the leaf is transformed into the stone, inclosing the one seed ; the tissue into the pulp, and the lower cuticle into the downy, blushing rind. The furrowed line on one side of the peach marks the union of the two edges of the carpellary leaf. The apple is a 5-carpelled fruit formed of the 5 united pistils. In its construction the 5 carpellary leaves are combined with the 5 calyx leaves. The upper surface of the former becomes the parchment-lining of the 5 cells of the core, and the tissues of them all grow into the luscious pulp. The orange is formed of twelve leaves, each transformed into a carpel, distinct in the pulp, but completely blended in the rind, while in the gooseberry the venation of the several leaves of which it is formed are still distinctly visible. The leaf is thus the rudiment, type, or pattern, whence develop every organ of the plant is developed, modified in color, shape, and structure to subserve, first, the special purposes in its own economy, and ultimately, the interests of the animal creation, and even man himself, " to whom the sweetness of the fruit and the beauty of the flower must have had reference in the gracious intuitions of Him who created them both."

XXX. THE LOCUST TREE.

Description.—This elegant and useful tree grows native in mountain forests, from the Ohio River southward, and is generally cultivated for timber, ornament or shade in nearly all the States.* In May and June, amid the general festivities of nature, the Locust displays her pendant clusters of white fragrant flowers, enlivening the dark green of the graceful foliage.

Analysis.—THE LEAF REGION.— The leaves are compound—of what form? How do they differ from the leaves of the Rose? What are the number, margin, outline and apex of the leaflets? They are connected with the rachis by a short stalk—*petiolules*. Are there any stipules? A pair of short, sharp *spines* occupy their place, especially in the younger shoots; we may call them *stipular spines*.

THE FLOWER REGION.—Compare the flower of the Locust with that of the Pea, and notice the differences in the calyx, banner, wings, keel, stamens, and especially the style (3). Is the inflorescence centrifugal or centripetal? It is a perfect example of a raceme. Are the stamens (1, 2) diadelphous? Compare specimens of the fruit (4) which have survived the Winter, or which ripen in September, with that of the Pea. Is the pod 1-carpelled or 2-carpelled? Has the seed 1 or 2 cotyledons? Any albumen?

Sensitiveness.—Note the tumid or fleshy bundles at the joints between the petiole and stem, and the petiolules and rachis. Have these any connection with the spontaneous

---

* The Locust Tree attains its greatest perfection in Kentucky and Tennessee, where it often rises to the height of 90 feet with a diameter of 4 feet. For strength and durability its timber is pre-eminent, and therefore largely employed in shipbuilding, railway ties, and fence posts. As a shade tree its beauty is often marred by the depredation of worms, which eat at the branches until they break and fall. When collected in groves it seems less liable to this evil, and grows with great rapidity, often reaching a height of 8 or 10 feet the first season.

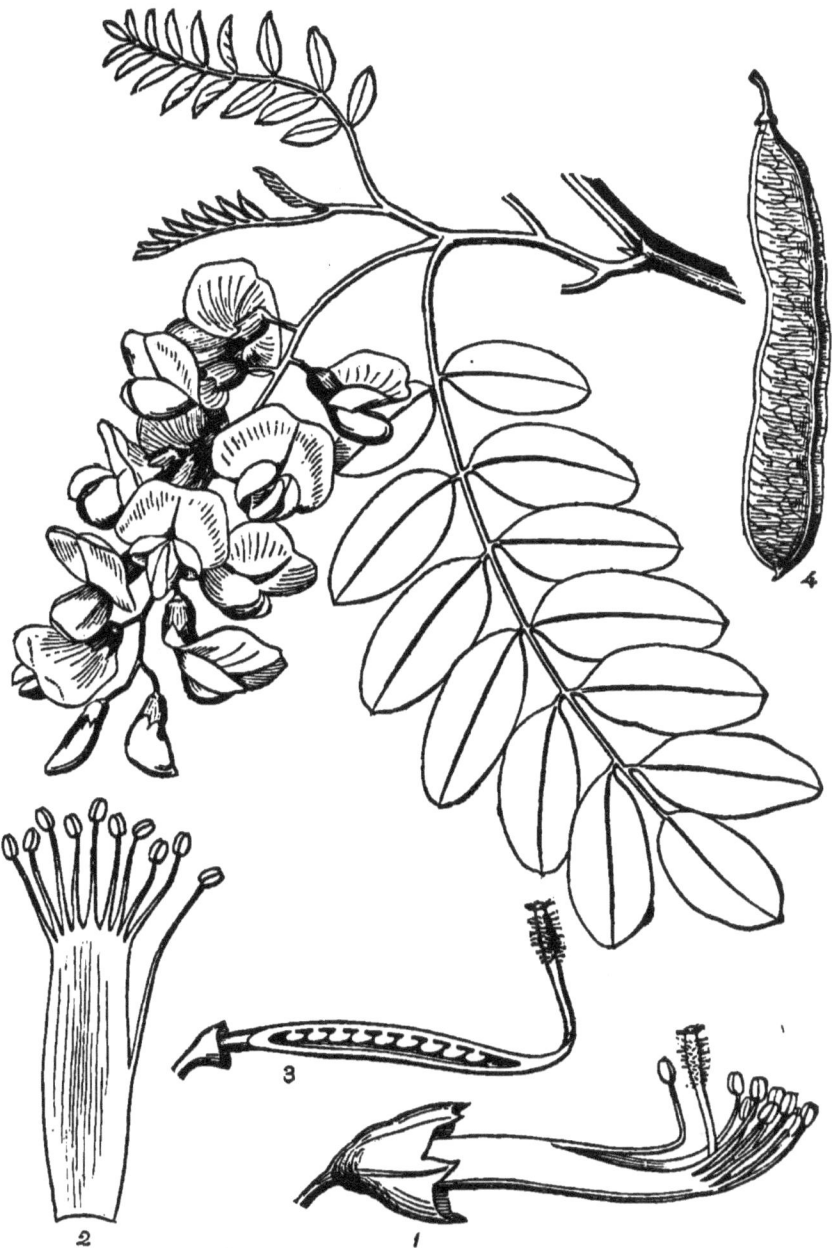

FIG. XXX.—Robínia Pseudacàcia: 1, the calyx, stamens and pistil, or the flower minus the corolla; 2, the stamens displayed; 3, the pistil with ovary dissected; 4, the fruit.

movements of the leaves for which the Locust is so remarkable? When, in securing your specimen, you grasped the branchlet, the leaves felt and, as it were, resented the violence. Did you not notice how they fell forward toward the branchlet, while every leaflet bent forward and upward until each met its fellow as if in sympathetic embrace?

Similar movements occur at evening with the departure of the sunlight. Then not a few leaves only, but the countless host on every branch bend, bow, and fold their leaflets face to face, and so sleep through the hours of darkness. When the dawn wakens the Robin to his song, it also wakes the Robinia, and her leaves with the advancing light slowly unfold to the sweet influences of the vital air.*

The True Sensitive Plant (*Mimòsa pùdica*) is native in tropical America from the Isthmus to Brazil. Its flowers are collected in roundish heads, its fruits are legumes of a peculiar pattern called *loments,* having joints between the seeds (5). The leaves are twice compounded (*digitate-pinnate*). When expanded, they are broad and showy, covering the plant with verdure. But at a touch of the finger, or the wing of a bee, they fold up and contract, one after another, so as almost to vanish from sight.† This results from a series of motions as follows: 1st, the numerous leaflets move upward and forward, twins meeting and together covering the pair next before them; 2d, the four divisions thus folded move toward one another as a fan

5, a loment.

---

* The leaves of our Wild Cassias, which open their yellow flowers in August, are also very sensitive, closing their numerous leaflets when touched.

† At Aspinwall, the traveler, first stepping from the car into a dense green patch of Mimosa, is confounded at seeing the whole patch disappear, leaving the ground almost bare, and again after a few minutes looking as verdant as ever!

closes; 3d, the whole leaf falls backward and downward by the joint at the base of the petiole.*

The Name, *Robínia,* was conferred in honor of John Robin, herbalist to Henry IV. of France, A. D. 1620. By his son the Locust was first cultivated in Europe in

5, Mimosa pudica; 6, Desmodium gyrans.

1640, under the popular name Acacia; hence its specific name, *R. pseudacàcia* (False Acacia). Two other species are native in southern forests and often seen in cultivation. (See *Bot. & Flor.*, p. 95.)

---

* The Moving Plant (*Desmòdium gyrans*) is another member of this great Order. It is native in India along the Ganges. In this country it is cultivated in the greenhouse. The leaves are pinnately trifoliate, consisting of a pair of very small leaflets placed a little below the large terminal oblong leaflet. These are wonderfully endowed with the power of spontaneous movements. Their motions are not occasioned by touch or irritation, but are voluntary and habitual. The small leaflets are more perceptibly active, moving steadily or fitfully, upward or downward or gyrating in circles, during the hours of sunshine. The large leaflet is quietly erect during the day, but slowly falls to a pendant position in the night.

Classification..—The vast Order of LEGUMINOSÆ, the Leguminous Plants, represented by the Pea, Locust, Cassia, and Mimosa, agree in having *alternate, stipulate, compound leaves, ovary simple, fruit a legume, and seeds without albumen.*

The **Leguminous Plants** number not less than 400 genera and 6500 species, 350 species being natives of the United States. The Order is remarkable for the beauty of its flowers, the variety and value of its products. Few, if any, are poisonous.

Among its food plants, are the Beans, Peas, Lentils, and Peanuts ; among its forage plants, the Clovers, Lucerns, and Carobs (called *husks* in Luke 15 : 16).

Of gums and balsams, we have Gum Arabic, Tolu, Senegal, Kino, Copaiva, Tragacanth ; of drugs and dyes, Indigo, Liquorice, Catechu, Senna, Logwood, Camwood, Brazil wood, and others innumerable.

The timber of the Locust tree, Laburnum, Dalbergia, and Itaka are highly prized in shipbuilding and cabinetwork.

Few fruits in flavor excel the Tamarind, and the powerful perfume of the Tonga Bean (*Dipterix odorata*) is well known.

Among its floral treasures, what element of beauty is lacking when we have the Wistaria, Golden Chain, Sweet Pea, the Acacias, Poinciana, and Clianthus ?

Scientific Terms.—Petiolules. Stipular spines.

XXXI. THE EVENING PRIMROSE.

Description.—A morning walk in June, through upland meadows, along fence-rows, and in sunny wastes generally, will be rewarded by the sight of the tall Evening Primrose, resplendent with its yellow flowers, which opened the night before. It is a biennial herb, 3 to 6 feet high, roughish, hairy, and leafy throughout.

Analysis.—The *Root* is axial, and usually biennial, like the Beet bearing only leaves during the first season, and storing away in the thick tuberous axis a surplus of nutritive matter to aid the larger growth of stem, flower and fruit the following year. By cultivation the tubers are improved in size and quality.*

The *Stem*, early the second year, rises erect and stout, 2–6 feet high, terete, hairy, simple at first, but at length widely branched like a little tree (hence often called the Tree Primrose). Sometimes the stem is rough, with short, bristly hairs.

The *Leaves* are many, closely ranged around the stem in an order called alternate, but easily seen to be in a spiral line running from right to left (see note, p. 193). Their outline is lanceolate, margin finely toothed in a manner called *repand-dentate*, like the border of a parasol. The lowest are petiolate, the upper sessile, and all pubescent.

Inflorescence centripetal, the lower buds opening first, forming a bracted spike which lengthens as the bloom advances upward, until, at length, there are at once fruit below, flowers in the midst, and buds at the top.

The *Flowers* are regular, symmetrical, 4-parted. The calyx is the remarkable feature. It consists of a long, slender tube adhering to the ovary below, expanding into 4 sepals at the top, where it also supports 4 broad yellow petals and 8 stamens.† The ovary is sessile, oblong, with a

---

* The tubers contain much nutriment, and before the discovery of the Potato were cultivated for food. Wine-bibbers ate them after dinner, as olives are eaten, supposing them to give greater relish to their potations.

† Fig. XXXI, 1, representing a vertical section of the flower of Evening Primrose, is worthy of careful study. It shows very plainly the nature of the adherent or superior calyx. Now it is understood that the floral organs all issue together from the torus (1)—the base of the flower. Then in this flower the sepals, petals, stamens and pistils are fused together into one body as far upward as the top of the ovary (o). At this point the style (which is compounded of 4) becomes free from the mass of the other organs, which continue in the form of a tube to the throat (e). Here the tube is resolved into its constituents, viz. the 4 sepals, the 4 petals, and the 8 stamens—all becoming free and distinct, and finally the style is also resolved into the 4 separate stigmas. In the related genus, Epilòbium, "the calyx tube is not prolonged above the ovary," but is resolved into distinct organs, all at once, at the summit of the ovary. Other genera, as Circæa, are intermediate between these two, having the calyx tube slightly prolonged.

Fig. XXXI.—Œnóthera biénnis: 1, vertical section of a flower; 2, 3, stamens; 4, pollen grains; 5, the 4 stigmas on 1 style; 6, a capsule, 4-valved; 7,— 4-celled; 8, a seed; 9, seed dissected; 10, the 2-lobed embryo.

long filiform style inclosed in the calyx tube and bearing at the summit 4 slender, spreading stigmas. The petals are contorted in æstivation. The anthers are versatile—fixed by the middle point. The pollen grains are angular, and loosely connected by spidery threads. After a night of bloom, the flower withers, breaks from the top of the ovary, and falls entire.*

The *Fruit* is an oblong, 4-sided, 4-celled capsule, filled with small seeds which have no albumen.

The Name.—*Œnóthera*, the title of this genus, comes from the Greek, meaning wine-hunter, from the notion that the roots cause a thirst for wine. The cuts represent *Œ. biénnis* (biennial), one of the many species.†

XXXII. LADY'S EARDROPS.

Description.—These floral gems are natives of the Andes from Mexico to Patagonia. They began to be known in Europe about A. D. 1780—in America, 1800; and are now universally cultivated. They are smooth, tender shrubs, requiring protection in our winters, and are propagated by slips and cuttings, as they seldom ripen their seeds.

Analysis.—THE LEAF REGION.—The root, as we grow

---

* The flowers open about seven in the evening, just at Summer twilight. The mode of expansion is very curious. The petals are held together at the summit by the hooked ends of the calyx. The segments of this flower-cup at first separate at the base, and the yellow petals may be seen peeping through these openings long before the flower is fully blown. The expansion is gradual until the petals are free from the confinement of the hooks, but when this is effected, the flower unfolds very quickly for a minute or two and then stops, after which it opens gradually, spreading itself out quite flat. The whole process occupies half an hour, and in some cases a little sudden noise is made as it jerks the topmost hooks asunder. It has been stated by Pursh and others that this plant, when in full flower, can be seen at a great distance, even in a dark night when all other objects are invisible, having a glow of bright white (see note, p. 85), as if its flowers were phosphorescent. There is evidently in this a reference to the visits of some night-flying moth adapted to suck its nectar in pay for scattering its pollen.

† Species 100, attaining their highest development in numbers and beauty W. of the Mississippi River. Their flowers are yellow, white, purple, 2', 3', or even 4' in diameter.

FUCHSIA. 129

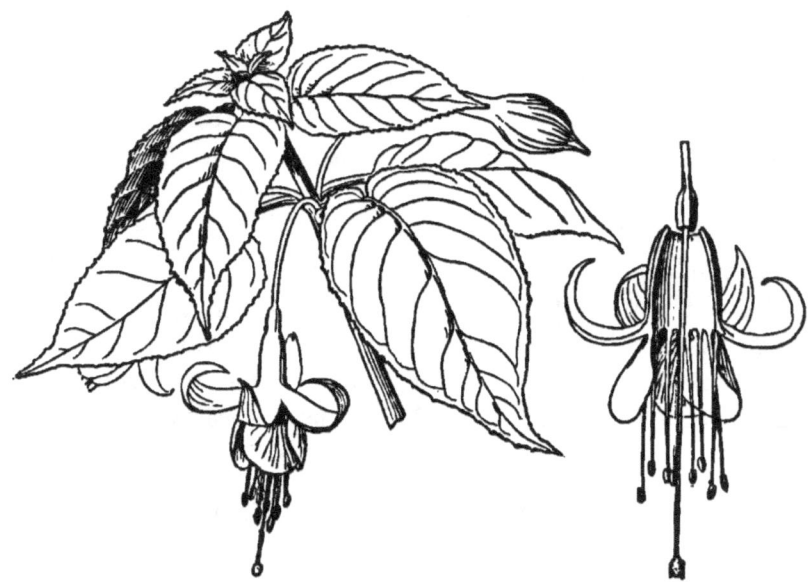

FIG. XXXII.—Fuchsia coccinia: 2. vertical section of a flower.

the plant, is represented by *adventitious* fibers issuing or "striking" here and there from the lower joints of the *slip* (severed branch) which remains as planted. The stem, although woody, is weak, slender, reclining, with smooth purplish bark and drooping branches. The leaves are neatly cut, smooth, pinni-veined with reddened veins, opposite, ovate, serrate and petiolate.

The *Flowers* issue either singly from the axils of the leaves, or clustered at the end of the branches, always drooping on slender peduncles. They are regular, perfect, symmetrical, 4-parted. The calyx is colored in contrast with the corolla, its tube inclosing, and adherent to, the ovary below, and opening into 4 lanceolate sepals above, bearing on its throat the 4 erect, convolute petals and the 8 long, *exserted* stamens. The style is free from the calyx tube, twice longer than the stamens, with a *capitate* (head-like; *caput*, head) stigma.* The pollen grains are angular and loosely webbed together as in Œnothera. Crowning the ovary within the calyx tube are 8 nectariferous glands.

The *Fruit* is a red berry, full of minute seeds, but seldom coming to maturity in this country.

The Name.—The genus *Fuchsia* was dedicated by Linnæus to Leonard Fuchs, a celebrated German botanist of the 16th century. *F. coccinia* (Scarlet Fuchsia), with flowers axillary, calyx bright red, from Chili, was the earliest known species. Others, now common, are *F. fulgens,* with flowers clustered, calyx tube longer than its lobes, which are often tinged with green, and ovate, somewhat heart-shaped leaves; *F. microphylla,* with small elliptical leaves, calyx funnel-shaped, and very sweet berries. But the species are much mixed by *hybridization.*† (*Bot. and Flor.*, p. 127.)

Classification.—Œnóthera and Fúchsia are members of the Order ONAGRACEÆ—the Onagrads. It will be seen that they coincide in the following points:

> Leaves simple, pinni-veined.
> Flowers perfect, symmetrical, regular.
> Calyx tubular, its lobes valvate in æstivation.
> Petals perigynous, convolute in æstivation.
> Stamens perigynous, once or twice as many as the sepals.
> Ovary inferior (adherent), 2–4 celled, with 1 style.
> Seeds anatropous, without albumen.

The Onagrads comprehend 22 genera, 450 species. They are chiefly natives of temperate climes, and specially numerous in America. They are of little importance to man, except for their beautiful and

---

* The arrangement of these organs seems nicely planned in favor of self-fertilization. But the falling pollen would seldom touch the *stigmatic end* of the stigma, where alone it would be effectual. Moreover the copious nectar implies that the help of some insect is still needed—some long-tongued moth or humming bird, probably, not found in this country. In New Zealand, a bird (Anthornis Melanura) is frequently seen with its head covered with the pollen of a native species of Fuchsia.

† Hybrids are artificially produced by transferring the pollen of one species to the stigmas of another, and planting the seeds which result.

showy flowers. *Zauschnèria* is a genus of handsome herbs, native of California, with flowers strikingly similar to those of Fuchsia. The Clarkias * of California are proverbially beautiful. The Willow Herb (Epilòbium augustifòlium), with its showy spike of blue-purple flowers, is a tall, familiar object in the New England wilds. The Enchanter's Nightshades (Circæa) are pretty little herbs of our damp woods, always welcome to the botanist for the charming simplicity of the flowers, being 2-parted throughout.

Scientific Terms in XXXI and XXXII: Adventitious. Capitate. Repand. Slip.

XXXIII. SWEET CICELY.

Description.—The Cicelys grow wild from Canada to Carolina, and westward to Oregon. Their favorite haunts are in damp, rocky woods. If there be a vein of water—a rivulet half hidden under decaying leaves, oozing along among stones and tangled roots, there will the Cicelys stand luxuriating in the rich mold in company with Toothroots, Trilliums, White Violets and other plants which bloom in May and June.

Analysis.—In the *Root*, Sweet Cicely possesses qualities which make it favorably known. It is perennial, enduring the frosts of many winters. It consists of a short body or *axis* soon dividing into several long, descending branches, all rather fleshy, sweet-scented when bruised, and with a spicy, anise-like flavor. It is esteemed in medicine as a tonic and expectorant.

The *Stem*, generally branching, arises 2 to 3 feet. The internodes are hollow, straight, uncommonly long, and minutely *pubescent*.

---

* Named for Capt. Clark of the famous Lewis & Clark's expedition, which made the first exploration of the Pacific Coast (1804). (See *Barnes's Hundred Years of American Independence*, p. 361.)

SWEET CICELY.

The *Leaves* are alternate, large, *decompound*—bi or triternate (the terminal divisions pinnatifid or pinnate), the radical one on a long petiole, the others nearly or quite sessile, leaflets thin, ovate, pointed, incisely toothed, sparingly pubescent. The petioles are peculiar, being flattened or winged below and so embracing or *sheathing* the stem.

The *Inflorescence* is in compound umbels, usually two together, terminating the stem and branches. Each compound umbel consists of 3 to 6 simple ones (*umbellets*), whose stalks are called *rays*. At the base of the umbel, are several

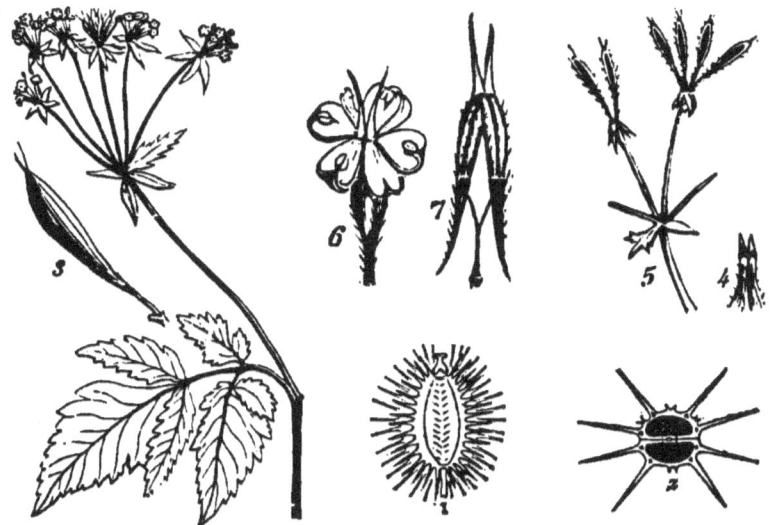

Fig. XXXIII.—Osmorhiza longistylis : 3, the flower ; 5, the fruit ; 8, a cremocarp ; 7, a cremocarp opening, on the carpophore ; 4, the short styles of O. brevistylis : 1, cremocarp of Carrot ; 2, the same in a cross-section.

(1—3) narrow bracts more or less leaf-like—an *involucre*. Also at the base of each umbellet, is a whorl of oblong bractlets bordered with hairs (*ciliate*)—the *involucel*.

The *Flowers* in each umbellet are about 5. Do you miss the calyx ? No sepals appear ; but under the flower, at the top of the pedicel, is a swelling which we may regard as the calyx-tube adhering to, and inclosing the ovary ; while the teeth (ends of the sepals), which we might expect to see as in the Apple flower, are *obsolete* or missing. The 5 small petals are conspicuous for their snowy whiteness. The point of each is abruptly *inflected* so as to make it appear notched (*emarginate*) at the end. There are 5 stamens, inflected like the petals. The two styles are prominent, slender, as long as the stamens, gradually enlarged at the base into the ovary, or rather into a *disk* which crowns it. The ovary is *inferior*, i. e., adherent to the calyx tube which incloses it, 2-carpeled, and 2-ovuled.

The *Fruit* of this Tribe of plants is of curious structure, and affords the best, often the only characters for distinguishing between the genera. In Cicely, its form is linear-oblong, with a tapering base—somewhat club-shaped, flattened on the sides, crowned with the 2 styles. It finally splits into 2 carpels displaying a *forked carpophore* (p. 75) on which each remains awhile suspended. The carpels are nearly terete, the face being narrow, and the back with 3 linear, *hispid* (with short stiff hairs) *ribs*. This form of fruit is called *cremocarp* (Gr. *kremao*, I hang, *karpos*, fruit).

The Name, *Osmorhiza* (root-scented), as well as the specific term, *O. longistylis* (long-styled), given to this plant, is characteristic. Another kind of Cicely (*O. brevistylis*, short-styled), growing in similar situations, will often be found and mistaken for this. In the former, the styles are slender and as long as the stamens : in the latter, conical and thrice shorter ; the leaflets more pointed and pinnatifid ; the bractlets long-pointed, and the root less agreeable in taste.

Scientific Terms.—Axis of root. Carpophore. Cremocarp. Decompound. Emarginate. Inferior ovary. Inflected petals. Involucre. Involucel. Obsolete. **Pubescent. Rays** of umbel. Sheathing petiole. Umbel. Umbellet.

XXXIV. GOLDEN ALEXANDERS.

Description.—The humid river-banks, the meadows behind them, and even the sunny hills above them, are frequently bedecked in June or May, with bright yellow umbels, which, with little discrimination, the country people call Golden Alexanders. We will suppose that our young botanists return from their morning rambles equipped with these plants complete—root, leaf, flower and fruit.

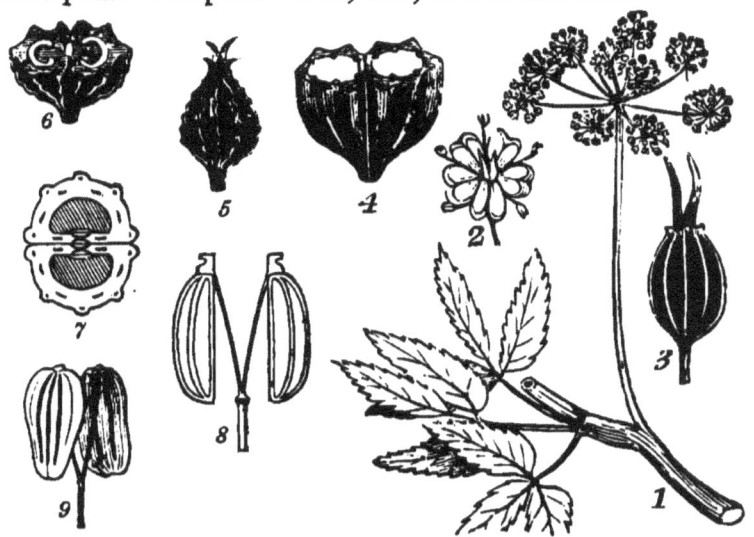

Fig. XXXIV.—Carum aùreum : 2, a flower ; 3, a fruit with its thread-like ribs and elongated styles ; 4, a cross-section of the same ; 5, a fruit of Conium : 6, its cross-section ; 7, cross-section of a fruit of Fennel ; 8, the same split into its merocarps suspended on the carpophore ; 9, a fruit of Parsnip, showing the vittæ, etc.

Analysis.—THE LEAF REGION.—After the lesson on the Cicely, the student will see in this plant striking analogies, with special differences. Both are to be carefully noted. The root is perennial, axial, branching, more woody than fleshy, from which annually arises a plant *glabrous* (smooth) and polished. The stems throughout are jointed, branching, with long, hollow internodes as in Cicely. The leaves are ternate and biternate, the lower on long petioles and sometimes pinnately 5-foliate, the very lowest being simple and cordate. The student will compare the leaflets with those of Cicely, and note their form of outline, base, apex, and margin. The petioles are sheathing and stem-clasping at the base, as in that plant.

THE FLOWER REGION.—The umbels are axillary and terminal.* Are they simple or compound ? Do you find any involucre and involucels ? Of what description? The flowers are 5-parted. Here also the calyx consists of a tube adhering to the ovary, with the limb or teeth *obsolete*. Each of the 5 yellow petals has its slender point inflexed, with the 5 stamens in like manner inflected. The ovary is inferior—placed below the flower and crowned by it, in consequence of being immersed in and adherent to the tubular calyx. The 2 styles are slender, longer than the ovary, and deciduous, for they are not seen on the full-grown fruit.

The *Fruit* is a *cremocarp* as in Osmorhiza, but with several remarkable differences. It is oval inclined to oblong, flattened on the sides. When the carpels separate, they show the forked carpophore between them. Each carpel has 5 conspicuous, equal, wavy ribs, 2 of which are marginal, i. e., on the border of the face or *commissure*. In each interval between the ribs is an *oil tube*—an oblong cell containing a fragrant oil. Botanists call these oil-tubes *vittæ*. None are found in the fruits of Osmorhiza.

---

* Plants in which the inflorescence is arranged in a cyme, corymb, &c., may be termed the "Social Flowers." Small flowers thus packed closely together are necessarily more attractive to insects than if they were scattered promiscuously over the plant. Besides, these groups of flowers are generally placed where they are not hidden by the leaves. So that one can but feel that this floral arrangement is not an accident, but designed for a purpose. Self-fertilization is guarded against in these masses of small flowers by the stamens ripening before the pistils. The former shed their pollen and wither before the latter have developed sufficiently to receive the pollen. Sir John Lubbock remarks that the honey in the flowers of this order is inaccessible to butterflies, whose proboscces are fitted for deep-throated flowers; but it is easily reached by other insects.

The Name in Latin is *Carum aùreum*. It is associated with Caraway (*Carum Carvi*) whose native country is Caria in Asia Minor; hence the name. The specific term, *aùreum*, means golden. Other plants called also Golden Alexanders, with yellow umbels in June, may perplex the student. One such, *C. cordàtum*, is smooth all over like *C. aùreum*, but its root-leaves are generally cordate and simple, and the stem-leaves never biternate.

Classification.— These examples introduce us to the great Order of the UMBELLIFERÆ—the Umbel-bearing Plants, characterized as we have seen by the following 7 traits:

Stems hollow. Inflorescence in umbels.
Leaves divided. Flowers pentandrous.
Petioles sheathing. Ovary inferior.
Fruit a cremocarp.

The Umbelworts.—The 152 genera of this Order, and probably also the 1500 species, are distinguished by as many varying forms of the cremocarp. Here the fruit is flattened on the sides; there, as in Parsnip, on the back, and in Coriander not flattened either way, but globular. Here the ribs are angular ridges; there they are winged; in Carrot they are each beset with a row of bristles. The ribs vary in number, from 3 to 9; so also the oil-tubes, being none in Cicely, 4 in Carum, 9 in Carrot, and 15 or more in Lovage. With a good microscope, the student will find these observations full of interest.

The Umbelworts are chiefly natives of the North Temperate Zone in both Continents, and the high mountains of the Tropics. Many of them are adapted to special uses. As food plants, we have the Carrot, Parsnip, Celery,* Parsley, Chervil. For aromatics and carminatives, we have the fruits of Anise, Caraway, Coriander, Dill, Cummin. As

---

* The action of light upon plants is well illustrated in the case of our Garden Celery. The stalks are blanched by heaping earth about them so as to exclude the sun; but not only is the formation of the green coloring matter (chlorophyl) thus prevented, but also, of the strong-odored if not poisonous substance which ordinarily renders this plant unwholesome.

drugs, Assafœtida,* Opoponax, Bdellium, Gum Galbanum, and the poisonous Conium, Cicuta, Fools-Parsley, etc., which all should know in order to avoid.

The Record.—For tablet and fig. of Cicuta, see Appendix.

Scientific Terms.—Commissure. Vittæ.

XXXV. THE MOUSE-EAR EVERLASTING.

Description.—These plants are among the earliest and oddest of the creations of Spring. On the sterile knolls of old pastures, and along the borders of the woods, you will find them already lifting their woolly heads when the grass first changes to green. Few plants are more unsightly, but being the heralds of returning Spring, the earliest representatives of the grandest of all the Orders, and moreover everywhere present, they make an undeniable claim upon our attention.

Analysis.—The *Root* is perennial, and produces upright flowering stems, together with prostrate runners or stolons like the Strawberry plant. All the herbage is whitened by a silky wool.

The *Leaves* are thickish, smoothish above when old, entire; the radical obovate or *oval-spatulate* (like an apothecary's *spatula*, or broader), petiolate; the cauline much smaller, linear-oblong, sessile. On the *stolons* (runners, p. 97), the upper leaves are the larger.

The *Stems* are about a span in height, and scape-like in consequence of the diminished upper leaves.

The *Flowers* are small, and collected in heads which are again assembled in clusters forming a dense terminal group. They are *dioecious*, that is, all *staminate* (♂) or *sterile* in

---

* Assafœtida is so much relished by the Brahmins of India that they term it "food for the gods."

THE MOUSE-EAR EVERLASTING.

FIG. XXXV.—Antennaria plantaginifolia: ♂, the sterile, ♀, the fertile plant; 1, a single floret; 2, a bristle of the pappus; 3, a ♀ floret; 4, a section of a ♀ head; 5, achenium with its pappus.

one plant, and all *pistillate* (♀) or *fertile* in another. The botanist should have both kinds in hand.

An *Involucre* consisting of many bracts or *scales*, surrounds each head of flowers. Here the scales are *scarious* or dry, white (or brown at the base), imbricated, the outer very woolly, the inner smooth, obtuse in the sterile heads, acute in the fertile.

The *minute flowers*, often called *florets*, stand crowded together on the *receptacle*—the expanded summit of the short peduncle. Here the receptacle is *naked*, i. e., bears no chaff among the florets. The ♀ florets show, first, an ovary at the base (inferior); 2d, a calyx (*pappus* *) consisting of about 20 fine white hairs crowning the ovary; 3d, a tubular corolla exceedingly slender, inclosing, 4th, a style protruding (*exserted*) from its summit. The ♂ show a slender abortive ovary at the base; a pappus of 20 club-shaped, knobby, white bristles; a tubular 5-toothed corolla inclosing 5 stamens whose brown anthers are united into a tube and *exserted*. The style is rarely seen.

Thus the fertile plants are known at sight by the longer, finer, whiter pappus not sprinkled with the brown dots of the anthers. The shorter, clubby bristles of the sterile pappus are curious objects under the microscope, but poorly contrived for wings.

The *Fruits* are each one-seeded—a sort of achenium. When ripe, they quit the receptacle, and, winged with their fine light pappus, are wafted away and scattered. For the abortive achenia, wings would be useless.

The Name of this plant is *Antennària plantaginifòlia;* the former suggested by the resemblance of the singular pappus to the antennæ of an insect; the latter by the likeness of the leaves to those of the Plantain.

---

* From the Latin *pappus*, an old man, a grandfather, alluding to the white hairs. Comparing this fruit with the *cremocarp* of Cicely (p. 131), it is evident that the ovary is *inferior*, i. e. the calyx tube adheres to the ovary, and the *limb* (sepals), if any, will seem to stand upon it, as the corolla does. But owing to its crowded condition in the dense heads, the sepals develop themselves in singular forms, usually split up into hairs or bristles, sometimes into 5 scales, as in Ageratum, sometimes into 2 teeth, as in Sunflower, and sometimes wholly obsolete, as in Mayweed. Again, the top of the ovary grows up into a neck elevating the pappus, as in Milkweed; or into a slender pedicel, as in Dandelion.

The Record.—Find in the Appendix a tablet and record of Antennaria, which will serve as a model for other plants of this order.

Scientific Terms.—Diœcious. Exserted. Fertile. Florets. Imbricated. Ovary Abortive. Pappus. Pistillate. Receptacle. Receptacle naked. Scales. Scarious. Scape-like. Spatulate. Staminate. Sterile.

XXXVI. THE ROBIN'S PLANTAIN.

Description.— The groves and orchards are already vocal with the song of the Robin when the meadows and copses are first bedecked with the blue rays of Robin's Plantain. In Florida beginning to flower in March, its bloom progresses northward to Virginia in April, to New York in May, and to Canada in June, coeval with Bulbous Crowfoot, Rue Anemone, and Hood-leaved Violet.

Analysis.—THE LEAF REGION.—Having collected an ample supply of specimens both with flowers in fresh bloom and others well advanced towards fruit, the student will answer inquiries like the following : What of the life and form of the root ? The quality of surface or the clothing of the plant? The form, stature, and attitude of the stem ? The position, arrangement, margin and venation of the leaves ? The leaves, as to outline, are not uniform. The radical are oblong-spatulate with the base narrowed toward a petiole and a few teeth above. The cauline are lanceolate-oblong, mostly entire, and with a broad-clasping base.

THE FLOWER REGION.— The *Inflorescence* is like Antennaria, but more open and *corymbous* (like a *corymb*, or a level-topped cluster). A few heads terminate the stem and branches. The first to flower is the one at the top of the stem, next that of the highest branch, and so on to the lower or outer. Thus the general inflorescence is centrifugal ; but regarding each head singly, centripetal.

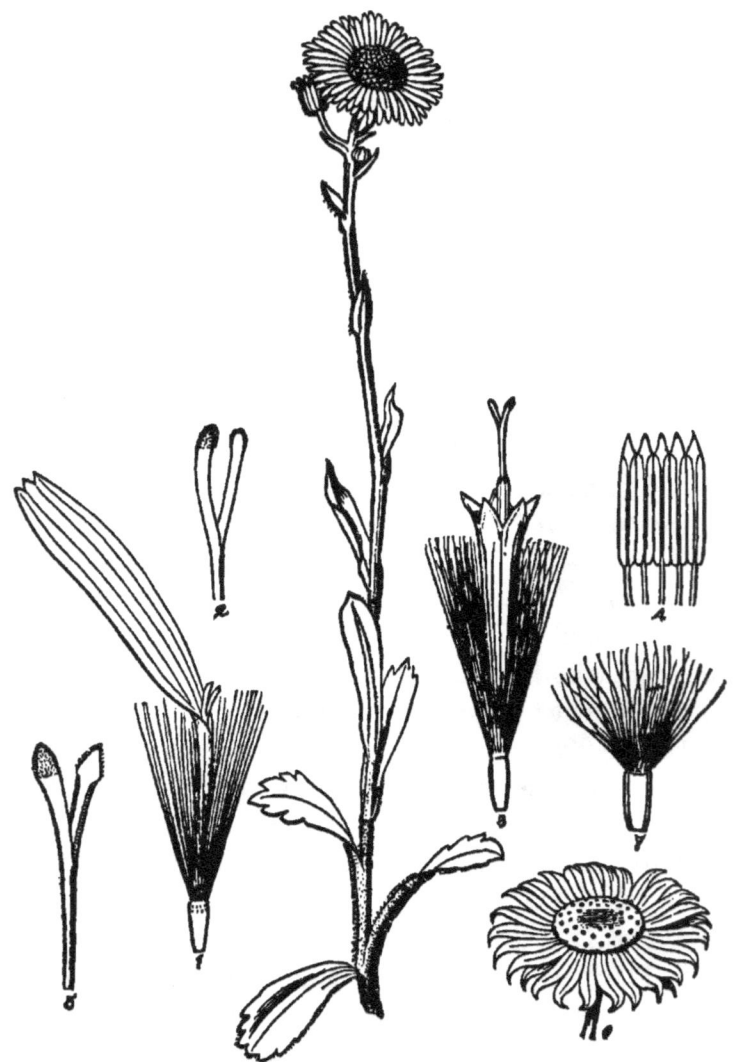

Fig. XXXVI.—Erígeron bellídifòlium: 1, a ray floret; 2, its style and stigmas; 3, a disk floret; 4, its stamens; 5, its pistil; 6, receptacle and involucre; 7, ripe fruit and pappus.

The *Involucre* consists of many nearly equal scales, green, linear, pointed, and all in one row, not imbricated (6). The receptacle (6) is flat, naked (no chaff among the florets).

The *Florets* are very many in each head and of two kinds. In the circumference stand the *florets of the ray*, surrounding the *florets of the disk*. The former (1) are distinguished by their *ligulate* corollas (*ligula*, a strap), called the *ray*. This form may be understood by comparing it with the *tubular* corollas of the disk (3). These have 5 teeth at the top, indicating that the tube is formed by the union of 5 petals.* There are also 5 (or at least 3) teeth at the tip of the ligules, and other marks indicating 5 united petals, not forming a tube, but a strap-shaped corolla—a split tube, lengthened and turned to one side. The rays in this species are of a bluish-purple color, and about 50 in number. The ray florets are pistillate (♀), the disk florets perfect (☿), and both are fertile. The style in all bears 2 manifest exserted stigmas. In the disk, 5 united anthers form a tube around the style; in the ray no anthers appear. In the figure, (4) displays the stamen tube as if unrolled; (5), the style with the 2 flattened obtuse stigmas; (2), the stigmas of the ray. The pappus is composed of many white, *scabrous* (rough) bristles encircling the corolla and crowning the (inferior) ovary.

The *Fruit* (7), a sort of achenium, is more properly a *cypsela*. The 2 stigmas indicate a 2-carpelled ovary. The cypsela, therefore, although 1-seeded, is the product of a double ovary. All other achenia (e. g., Ranunculus) come from simple ovaries.

The Name of this plant, *Erigeron bellidifòlium*, is singularly descriptive. The generic title signifies "hoary in Spring" (*ĕr*, spring, *geron*, an old man); and the specific, "daisy-leaved" (*Bellis*, Daisy, *folium*, leaf).

---

* The term *gamopetalous* (*gamos*, union) is applied to all flowers with united petals. The corresponding term *polypetalous* designates those having the petals distinct, as in the plants heretofore described.

Scientific Terms.—Corymb. Corymbous. Cypsela. Florets of the ray. Florets of the disk. Gamopetalous. Ligulate corolla. Polypetalous. Rays. Scabrous. Tubular corolla.

XXXVII. THE DANDELION.

Dear common flower, that growest beside the way,
Fringing the dusty road with harmless gold,
'Tis the Spring's largess which she scatters now.
                                                    LOWELL.

Description.—There are animals which shun the savage haunts of the wilderness, and with determined choice seek the habitations of man. So there are plants, foreigners mostly, such as the Plantain, Pigweed, and Dandelion, which flourish only or chiefly around human dwellings. Early and late, in Spring, Summer and Autumn, the golden discs of the Dandelion develop from the manipulated soil of the gardens, fields, and fence-rows. Other plants we may value for their rarity; but this delights us for its very commonness, and the associations of childhood which linger about it.*

Analysis.—THE LEAF REGION.—Here we have an *acaulescent* plant—a plant with no visible proper stem. The leaves and flower-stalks rise directly from the top of the strong, axial, fleshy, perennial root. A milky white juice pervades the whole plant, exuding from the root, leaves, flowers, wherever bruised or broken. This juice contains caoutchouc, but no opium. The leaves differ in pattern from any hitherto described. All are radical, and oblong in their general outline, with the margins cut into prominent lobes and teeth which are inclined backward—a form called *runcinate* (re-uncinate, or hooked backward).

---

* Besides the uses of the Dandelion for the bee, butterfly, and childhood, and the pleasant memories it brings to age, it serves other purposes. The young leaves when blanched are esteemed in France as an excellent salad. The green growing leaves are used generally as a pot-herb. The root is a valuable remedial agent.

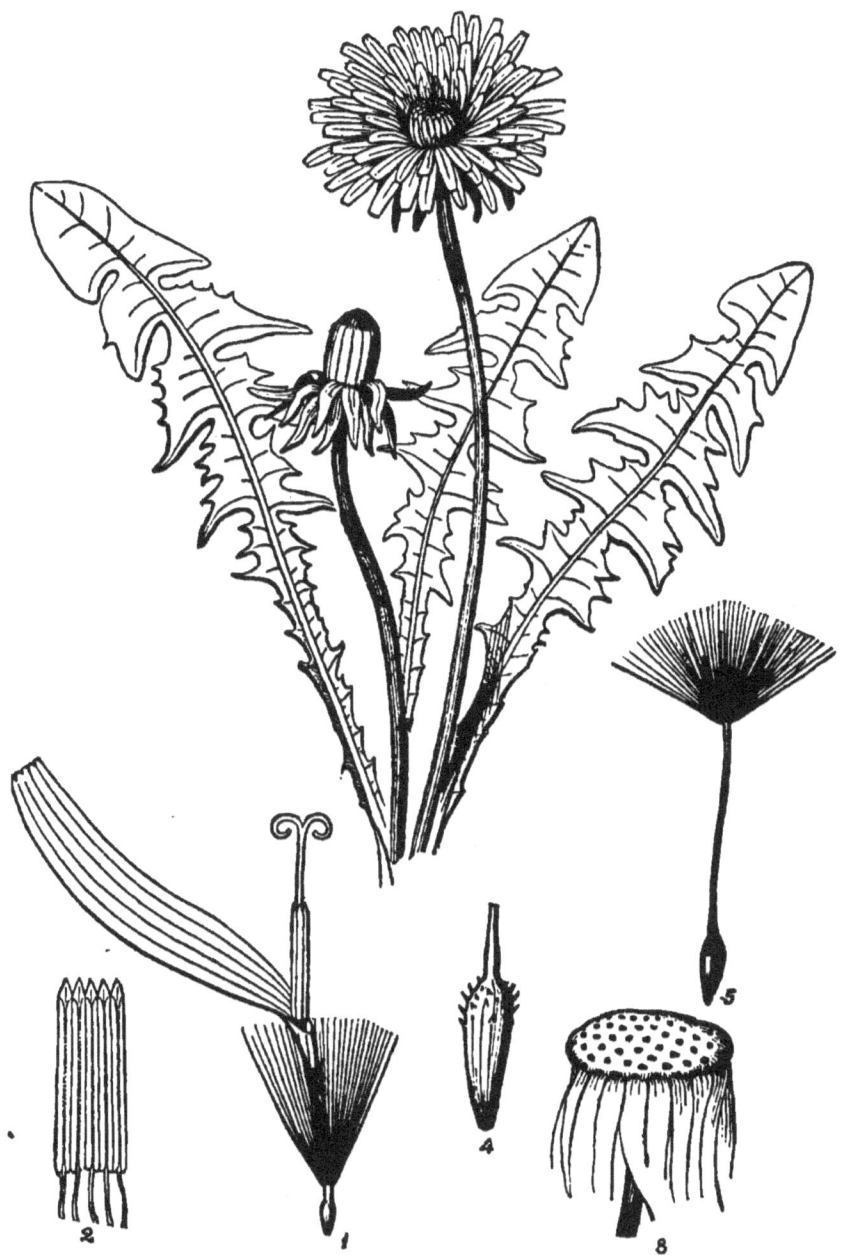

Fig. XXXVII.—Taráxacum Dens-leónis: 1, a floret; 2, the stamens unrolled; 3, the receptacle, and involucre; 4, a fruit (cypsela); 5, a fruit with its pappus.

THE FLOWER REGION.—The inflorescence is also radical. From the crown of the root several naked, hollow scapes arise, each bearing an involucrate head of flowers analogous to, yet strikingly different from, those of Antennaria or Erigeron. In the former, the heads are *discoid*, being wholly destitute of rays or ligulate corollas; in the latter *radiate*, having the outer row of florets ligulate. In Dandelion they are *radiant*—with all the florets ligulate.

The *Involucre* is said to be double, consisting of 2 rows of scales, the outer ones shorter and reflexed, the inner, linear and erect. The receptacle (3) is quite naked of chaff.*

The *Florets* are all fertile and perfect—each consisting (1) of an oblong ovary crowned with a yellow, ligulate corolla and a pappus of soft white bristles. The ligule is 5-toothed, indicating 5 united petals. The 5 anthers form a tube inclosing the style, which divides at the top in 2 spreading or revolute stigmas. The anther tube is represented in the cut (2) as if unrolled.†

The *Fruit*. After flowering, the involucre closes upon the withering corollas while the fruit is growing. The tips of the ovaries grow into slender *beaks* raising the pappus, while the scape lengthens, elevating the whole head. At length, when all is ripe, the involucre again opens, the pappus expands into an airy balloon, and soon the cypselas (4 and 5), thus admirably fledged, are borne away on the wind and scattered far and wide. But this is not the end of

---

* In Sunflower, Coreopsis, and other plants of this Order, the receptacle bears with each floret a bractlet (called a *pale* or *chaff*). Hence "*receptacle chaffy*" is the counterpart of "receptacle naked."

† In fair weather the florets are expanded and very conspicuous to insects. In rain and by night they are closed, protecting the nectar from waste. The nectar is abundant, rising high in the tubes of the florets and accessible to numerous insects. Müller observed the visits of 93 species. It is scarcely possible that the stigmas should escape pollenization in this way; but to make sure of it, they continue to recoil until they reach the pollen for themselves. "The brightness of its color, the quantity of its honey, the habit of closing in unfavorable weather, and the power of self-fertilization, go far to explain the great abundance of the Dandelion."—*Sir J. Lubbock.*

providential care. The cypsela (4) is pointed and bearded so that when it alights, its pappus still moving to and fro, it works its way into the ground and thus plants itself.*

The Name, Dandelion, is a corruption of the French *dent-de-lion,* from a fancied resemblance of its jagged leaves to the teeth of a lion. The scientific name, *Taráxacum Dens-leònis* (*taraxacum,* disturbance, *dens-leonis,* lion's tooth) refers to this common notion, and its medicinal effect.

Classification.—This plant, with the two foregoing, introduces us to the great Order of the Asterworts, called COMPOSITÆ as the flowers are apparently *compound.* They agree with one another and with the whole Order in these seven characteristics :

1. Flowers collected in involucrate heads.
2. Calyx limb (if any) a dry pappus crowning the ovary.
3. Corolla of 5 united petals (gamopetalous).
4. Stamens 5, united by their anthers into a tube.
5. Stigmas 2, with their styles consolidated into one.
6. Ovary inferior, 1-ovuled, a cypsela in fruit.
7. Seed with no albumen.

The Asterworts embrace 766 genera and 9000 species, growing in all climates and countries, amounting to about one-tenth of the Flowering Plants of the Globe. Over 600 species are natives of the

---

* Thus the Dandelion enters the great "struggle for existence" with seeming advantages, but none too many. Its rivals are a legion, each in its own way armed for the strife—a contest more active than ever was waged on any human battle-field, renewed every Spring time in the bosom of the quiet woodland and peaceful meadow. The ground is densely packed with seeds which were strown the previous Autumn, or have been lying dormant, abiding their time, perhaps for years. There is room for only one seed to develop in a spot where there are hundreds of candidates. The sunshine and heat stimulate them to germination, and then begins the fierce struggle for survival—a contest that knows no pause or cessation until the fittest have conquered and the rest have succumbed. It is literally a death-struggle. No pity is shown for the weak, no regard for the beautiful.

Nowhere is this life-struggle so reckless as amid the exuberance of a tropical forest. "There," says Orton, "the dense dome of green overhead is supported by crowded columns, often branchless for 80 feet. Individual struggles with individual, and species with species, to monopolize the air, the sun, and the soil. In their efforts to spread their roots, some of the weaker sort, unable to find a footing, climb a powerful neighbor and let their roots dangle in the air, while many a full-grown tree has been lifted up, as it were, in the strife, and now stands on the ends of its stilt-like roots so that a man may walk under the trunk between them."

United States. Conspicuous among them are the autumnal hosts of blue and white Asters and yellow Goldenrods (*Solidágo*), the troops of Sunflowers (*Heliánthus*) and the armies of Thistles (*Cnicus*). Our Composites are nearly all herbs ; in Chili, they are mostly bushes ; in the Island of St. Helena, they are trees.

Compared with its vast extent, the useful products of this Order are few and unimportant. Lettuce is the herbage of *Lactùca sativa*. Salsify is the root of *Tragopógon porrifólius*. Chickory, used with Coffee, is the roasted root of *Cichòrium Intybus*. Saffron, a yellow dye, is the dried flowers of *Cárthamus tinctòrius*. Camomile (*Anthémis nóbilis*), Elecampane (*Inùla*), Arnica (*A. montána*), are popular remedies. The well-known *Persian Insect Powder* is the dried and pulverized heads of *Chrysánthemum ròseum*. Wormwood, used in making the French liquor absinthe, is a species of *Artemisia*.

In the flower garden this great Order is also well represented by the showy China Asters, Zinnias, and Dahlias, which sport into varieties infinite. The Feverfue (*Parthènium*), the Marigolds (*Tagetes*) and Coreopsis, are old favorites. The Everlastings or Immortelles are becoming common. Last, but not least, the lovely *Chrysánthema*, in purple, yellow, and glowing white, when all other leaves are falling and other flowers are dead.*

Scientific Terms.—Acaulescent. Achenium or Cypsela beaked. Heads discoid. Heads radiant. Heads radiate. Receptacle chaffy. Receptacle naked. Runcinate.

XXXVIII. THE CHECKERBERRY.

Description.—The many names of this little plant, as with the Garden Violet, is a proof of its popularity. In different places it is known as Checkerberry, Boxberry, Teaberry, Ivory Plum, Partridge-berry, Wintergreen.† The

---

* "And it is told in stories old that this fair blossom first
On that blest morn, when Christ was born, into *white* beauty burst.
Perhaps—ah ! well, we cannot tell if truly it be so ;
I but repeat the legend sweet, and only this I know—
That in the prime of Christmas-time the Christ's sweet flowers blow."

† We adopt the first-mentioned name, for so it was known to our childhood. But since these appellations are merely local, and some of them are equally applied to other plants, the necessity of an invariable scientific name is manifest.

Checkerberry grows in old woods and pastures, particularly where Pines and Hemlocks have abounded, always avoiding alluvial or limestone soil. In Winter and early Spring, it appears arrayed in the dark evergreen leaves and bright red berries of the preceding year. In April and May, it puts forth new leaves which are of a livelier green, and tender,

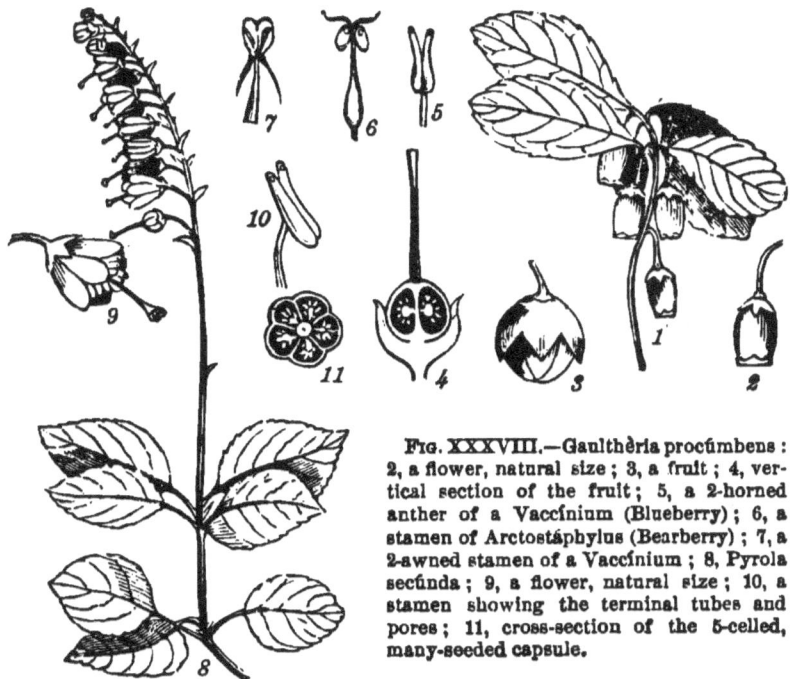

FIG. XXXVIII.—Gaulthèria procúmbens : 2, a flower, natural size; 3, a fruit; 4, vertical section of the fruit; 5, a 2-horned anther of a Vaccínium (Blueberry); 6, a stamen of Arctostáphylus (Bearberry); 7, a 2-awned stamen of a Vaccínium; 8, Pyrola secúnda; 9, a flower, natural size; 10, a stamen showing the terminal tubes and pores; 11, cross-section of the 5-celled, many-seeded capsule.

with their well-known spicy fragrance and taste. In May, June, or July, according to the locality, you will find it in flower, and in October its fruit will again be perfected.

Analysis.—THE LEAF REGION.—With complete specimens in hand, the learner will carefully distinguish between root and stem. Is it ⊙ or ♃ ? The stem proper, or the main stem, is a prostrate creeper generally concealed. At intervals it sends up branches 2' or 3' high, which the learner at first sight might take for distinct, independent stems. The leaves—define their venation, form of outline, qualities of surface, texture, taste, &c.

THE FLOWER REGION.—The *Flowers* present us with a new pattern. The white wax-like corolla is a short tube, 5-toothed and slightly contracted at the mouth—a form called *urceolate* (urn-shaped). Note the position, attitude, and length of the flowers and their peduncles. Compare the calyx with the corolla; note their difference in life, form, size, and fail not to observe the 2 bractlets subtending all. Note the number of the stamens, and the form of the anther. Each of its 2 cells bears an awn at its tip, and a terminal pore for the discharge of its pollen.

The ovary—is it superior? How many cells has it? How many ovules in each cell? How many styles? The microscope will reveal all this.

The *Fruit*. If the pupil has been fortunate in securing fruit, relic of the former year, new surprises await him in its analysis. As a whole it is globular. At the top appears a little globe within a globe, surrounded by 5 large teeth. Now with a sharp blade divide the fruit perpendicularly and study the section. There is a 5-celled capsule enveloped in (but free from) the enlarged fleshy calyx which contains the pulpy portion of the berry.

The Name.—This plant was first noticed in Canada by Dr. Gaulthier of Quebec. In his honor it received the generic name *Gaulthèria*, conferred by Prof. Kalm, of Sweden. Its specific name, *G. procúmbens*, alludes to its habit of growth. Another species, *G. Shallon*, a bush with similar fruit, but *black*, and the delight of the bears, grows in Oregon.

Scientific Terms.—Urceolate corolla.

XXXIX. THE PYROLAS.

Description.—There are five or six species of these elegant plants growing in the woods of the Northern States, Canada, and southward along the mountains. One or all of them may fall in the way of the collector, the flowers in June or July, the fruit in September. The dry stalks of the last season with empty pods (better than none) should be collected with the flowers.

Analysis.—We now adopt a new

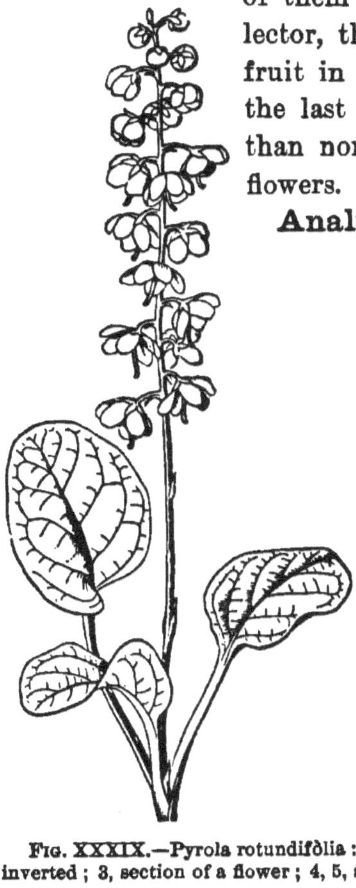

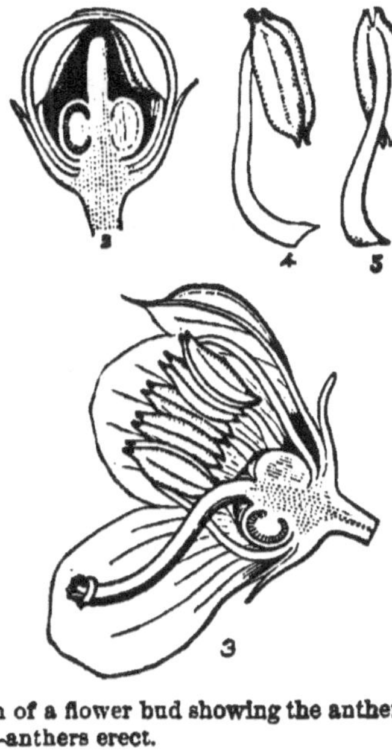

FIG. XXXIX.—Pyrola rotundifòlia: 2, section of a flower bud showing the anthers inverted; 3, section of a flower; 4, 5, stamens—anthers erect.

method of analysis. We direct attention, 1st, to such characters as apply equally to all the species of a genus, i. e., the generic characters. These are mainly but not entirely found in the flowers and fruit; 2d, to those which apply to one species only and serve to distinguish it from all the other members of the genus. They are taken from any part, but chiefly from the leaf region.

1. GENERIC CHARACTERS. — The Pyrolas are smooth, nearly acaulescent, perennial herbs. Both *roots* and *stem* are mostly subterranean. The former are brown fibers springing here and there from the joints of the stem. An underground shoot or runner arising from the base of the last year's plant, becomes the stem of this year's plant, and so on.

The *Leaves* are entire, petiolate, and nearly radical.

The *Inflorescence* is a scape with a few bracts, and a simple raceme.

The *Flowers* are complete, 5-parted, symmetrical, one-colored, nodding. The sepals are 5, united at the base, persistent. The petals are 5, larger, concave, *converging* (not wide-spread), scarcely united at the base, deciduous. The 10 stamens are peculiar in form and behavior. The large oblong anthers (4, 5) are attached to the top of the filament near their own apex, where they open by 2 (or 4) pores. In the bud (2) they are seen inverted, but become erect with their pores upward as the flower expands. A vertical section (3) displays the structure and arrangement of the floral organs.

The Style is one, compounded of 5 united, with 5 stigmas at the top. The superior ovary becomes in fruit a globous-depressed, 5-lobed, 5-celled capsule, opening upward from the bottom by 5 valves. The seeds are innumerable and very minute.

2. SPECIFIC CHARACTERS.—We have the portrait of a common species, and assume that the learner has specimens before him. The few leaves are quite radical, thick and

shining, *orbicular* (round) or round-ovate, shorter than their dilated petioles. The scape is 6–12' high, 6–12-flowered, bracted. The calyx lobes ovate; the petals round-obovate, nearly white. Style *clavate*, twice bent downward, longer than the petals which are thrice longer than the sepals.

The Name, in Latin as in English, is *Pyrola*, a diminutive of Pyrus, the Pear-tree; because of the resemblance of the leaves, whence it is also called, Pear Wintergreen. The species here figured and described is *P. rotundifolia*, the round-leaved.

Another species, *P. elliptica*, has elliptical and oval leaves, thinner in texture, scape bractless, and sepals very short. The learner may also have found *P. secunda* (Fig. XXXVIII, 8), which has the flowers of its raceme all turned one side, a straight style, serrulate leaves not all radical, and other differences which are easily noted.*

Scientific Terms.—Clavate. Converging petals. Generic characters. Orbicular. Specific characters. Vertical section.

XL. PRINCE'S PINE.

Description.—In the same woodlands where the Pyrolas grow, or in the drier portions of them, you may also detect the Prince's Pine, or, as it is called in the Indian tongue, Pipsissewa. The affinities of this comely plant with the

---

* In the States E. of the Mississippi River, 6 species of Pyrola are known, and analytically distinguished in Wood's Object Lessons as follows:
§ Stamens ascending, style declined and curved...*a*.
§ Stamens and style straight and erect...Nos. 5, 6.
    a Leaves thick and shining. Flowers white or rose-colored, Nos. 1, 2.
    a Leaves green, not shining. Flowers greenish-white, Nos. 3, 4.
1 P. rotundifolia. *Round-leaved P.* Leaves orbicular. Mostly white petals.
2 P. asarifolia. *Heart-leaved P.* Leaves round-cordate. Rose-colored petals.
3 P. elliptica. *Pear-leaved P.* Leaves large, thin, elliptical. Scape bractless.
4 P. chlorantha. *Green-fl. P.* Leaves small, thick, roundish, shorter than petioles.
5 P. secunda. *One-sided P.* Raceme with the green-white flowers all on one side.
6 P. minor. *Lesser P.* Raceme spike-form, with small, globular, white flowers.

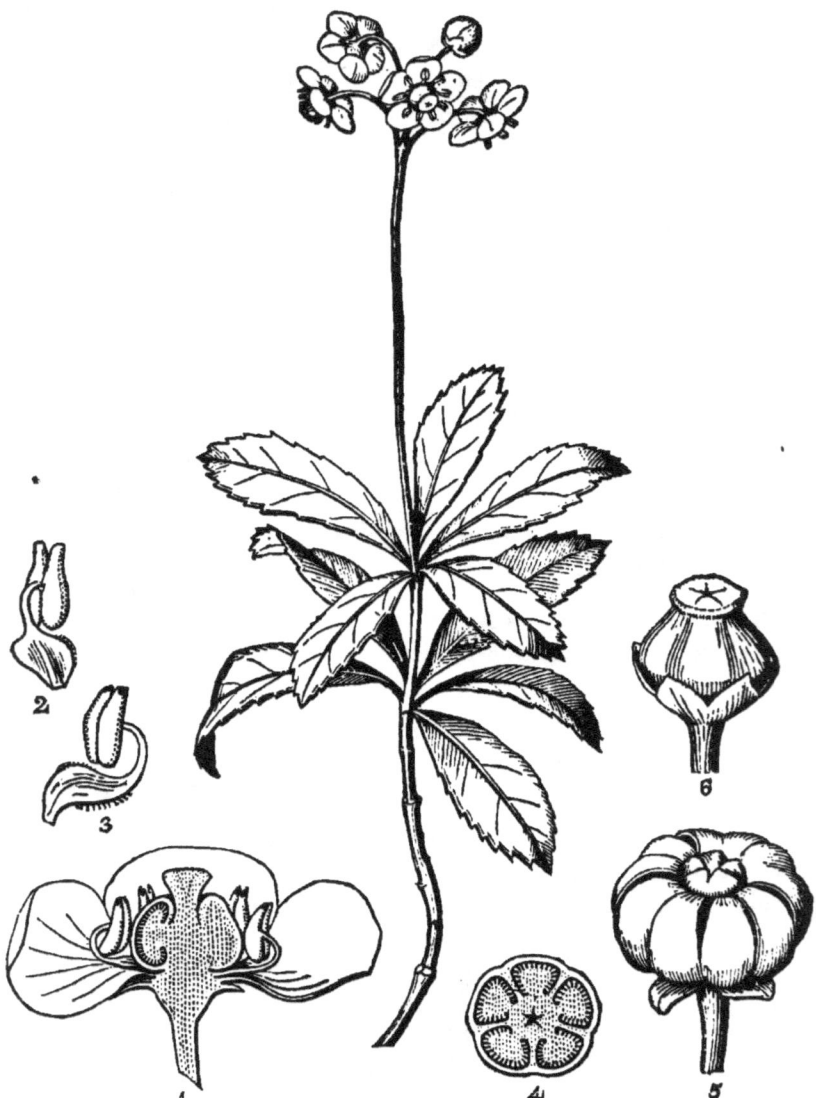

FIG. XL.—Chimáphila umbellàta: 1, section of a flower; 2, 3, stamens; 4, cross-section of ovary; 6, the ovary; 5, capsule opening by chinks above.

Pyrolas is evident at sight. Its study will therefore be a comparative analysis, in which both its resemblances and differences will appear.

Analysis.— 1. GENERIC CHARACTERS.—The Prince's Pines are small, *suffruticous* (*sub*, under, or partly, *frutex*, a shrub ; i. e., half-shrubby) plants. Their stems originate from long subterranean runners like the Pyrolas, with leaves evergreen, thick, shining, verticillate (whorled) or scattered in the midst of the stem. The inflorescence is a terminal umbel on a long peduncle, with flowers flesh-colored, 5-parted. The calyx is 5-lobed, and the corolla of 5 concave, orbicular, wide-spread petals. There are ten 2-horned anthers, opening by 2 terminal pores ; filaments (2, 3) broad in the middle ; style (6) very short ; stigma broad, disk-form. The capsule (5, 4) is depressed, globular, 5-celled, 5-valved, opening from the top.

7, Chimáphila maculàta.

2. SPECIFIC CHARACTERS.—The specimens in hand may be of the kind commonly known at the North as Pipsissewa (see Fig. XL) and esteemed for its tonic and diuretic properties. This plant stands 6–10' high on a base curving upward. The leaves are in 2 or 3 whorls of 3s and 5s, oblanceolate, narrowed to the base, sharply serrate, uniformly dark-green. The peduncle is 2–4', and sustains an umbel of 4–7 flowers.

The Name, *Chimáphila* (winter-loving) *umbelláta* (umbellate) is appropriate to the habit of Prince's Pine, it being an umbel-bearing evergreen. A second species—*C. maculàta* (Spotted Chimáphila, Fig. XL, 7) grows in similar localities, especially southward. It is known by the white variegations of the lanceolate, remotely serrate leaves. Sooner or later the diligent collector is sure to find it and record its analysis.

XLI. THE KALMIAS, OR AMERICAN LAURELS.

Description.—In the woods of the Atlantic States from Maine to Georgia and westward to Wisconsin and Kentucky, grow the American Laurels, adorned in the months of May and June in their magnificent bloom. Five species are known to the botanist, flowering simultaneously, and therefore, possibly, all or several in the box of the collector. Let us first notice their points of agreement, that is, their—

1. GENERIC CHARACTERS. These are evergreen shrubs, with coriaceous, entire leaves, with raceme-like *corymbs* (level-topped clusters) of showy white or red flowers, all 5-parted, gamopetalous and complete. The 5 sepals cohere only at base, the 5 petals are united quite to the top into a saucer-shaped, 5-lobed corolla larger than the calyx, having 10 pits or sacks in which the 10 anthers are lodged. The filaments are long, slender, elastic and recurved. Style 1, slender. Fruit a globular capsule, 5-celled, ∞-seeded.

The Name.—Plants possessing these attributes constitute a genus named *Kálmia*, in honor of *Kalm*, a Swedish botanist who traveled in America about 1750.

2. SPECIFIC CHARACTERS of the Broad-leaved Kalmia (*K. latifòlia*), often called the Calico Bush. It is a shrub with crooked stems and branches, 5–15 feet high, with leaves mostly alternate, smooth, bright green on both sides, ellip-

156   THE KALMIAS.

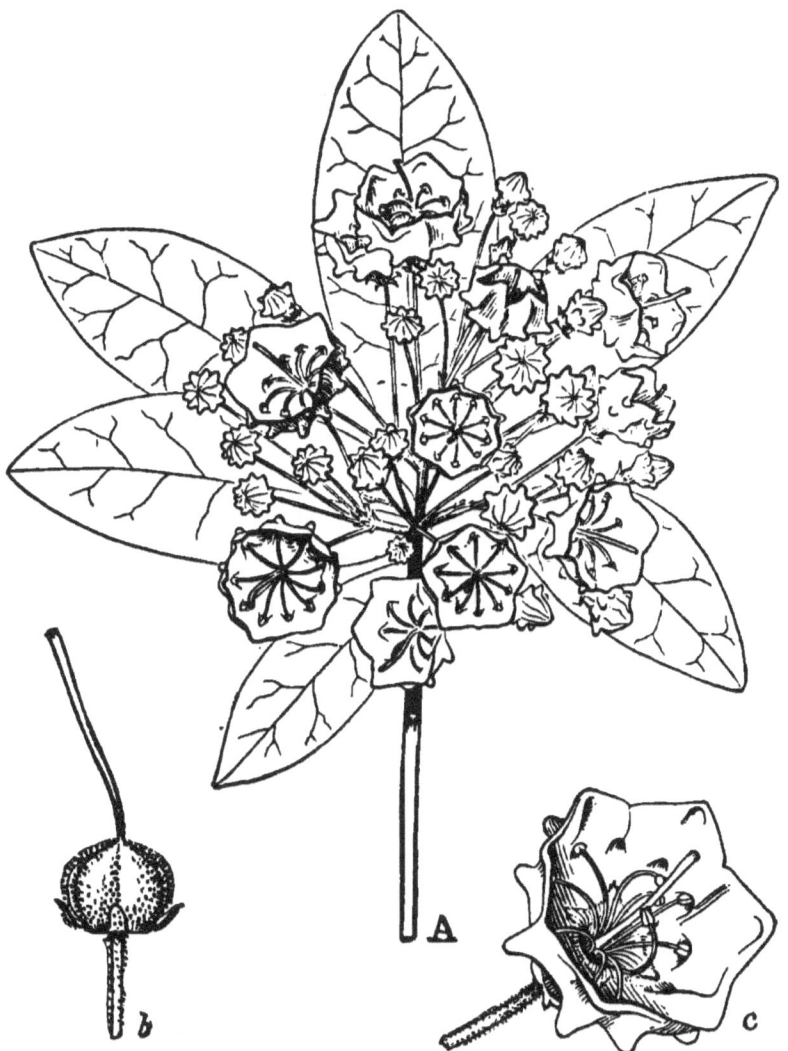

Fig. XLI.—Kálmia látifòlia: *c*, a flower natural size; *b*, a growing ovary, with its style.

tical, acute at each end, supported on short petioles. Its flowers are in large terminal corymbs, viscid-pubescent, white varying to rose-color.

The other species are shrublets 1–3 feet in height. The

learner will recognize their generic characters in the sketch already given, and write in his Plant Record the distinguishing specific characters of each as he finds them. Their names are as follows :

K. angustifòlia, the narrow-leaved, called Sheep-poison, with opposite leaves and lateral umbels.
K. cuneàta, the wedge-shaped-leaved, growing South.
K. glauca, the glaucous or sea-green ; with 2-edged twigs.
K. hirsùta, the hairy ; with very small leaves. South.

Pollenization. The curious action of the stamens in these plants is worthy of special attention. When the flower first opens, the stamens are confined with their anthers in as many little pockets of the corolla, and consequently the elastic filaments bend backward away from the style, which stands erect in their midst. The anther containing the pollen grains has, like the Pyrolas, 2 porous openings at the top. A touch or sudden jar will liberate the anthers, when they instantly rebound against the style, discharging their pollen toward the stigma.*

Classification.—The genus Kalmia, with the three preceding, represent the large and interesting Order of the Heathworts, or ERICACEÆ.† They are not homogeneous, like

---

* It has been observed that the stamens do not spontaneously free themselves, but await some external force, as a gust of wind, a falling twig, or rain-drop. But the special agent in this service is the bee in quest of honey. The rustling of its wings, the thrusting its proboscis into the cavity at the base of the stamens where the nectar is secreted, sets them free. In this case the pollen shot from the rebounding stamens will be discharged upon the body of the insect, and thus carried to the stigma of the next flower which it may visit.

The thoughtful student will here inquire, " Why must the pollen be lodged upon the stigma at all?"—a question which we are preparing to answer.

† The Order takes its name from its principal genus, Erìca, the Heaths or Heathers, a genus of not fewer than 400 species of delicate evergreen shrubs, with small narrow leaves and 4-parted gamopetalous flowers, natives of Europe and S. Africa. In Scotland, the luxuriant Heather is a characteristic feature of the landscape. It covers wide tracts of country so closely as to prevent all other vegetation, and often grows high enough to hide a man standing erect. Different species are the badges of different families, and a plant that is so serviceable is well worthy of being a Highland badge. Many a mountaineer sleeps on a couch of Heather boughs ; makes his cabin of Heather and a mortar of straw and earth ; thatches his roof with Heather, which he binds down with a rope of twisted Heather ; and burns for his only fuel the

the Compositæ, yet nearly all the genera agree in the following seven characters.

>Leaves simple, without stipules.
>Flowers perfect, complete, regular.
>Petals 4, 5, rarely more, united or not.
>Stamens as many or twice as many, free and distinct.
>Anthers 2-celled, opening by 2 terminal pores.
>Style 1 with a 4–10-celled ovary.
>Embryo small, in fleshy albumen.

The Heathworts comprehend 61 genera, 1330 species, chiefly natives of S. Africa, where they cover vast tracts of country, and America, both N. and S. Some of them are the most beautiful of plants, as the Azalias, Rhododendrons, and Heaths (Erica).

Our Blueberries, so delicious and healthful, are the fruit of the various species of *Vaccinium*. Our Whortleberries or Huckleberries, of *Gaylussácia* (dedicated to the French chemist, *Guy Lussac*). Our Cranberries, of *Oxycóccus*.

The Oil of Wintergreen is distilled from the young leaves of *Gaulthèria procúmbens*. The diuretic medicine, Uva-ursi, is the leaves of *Arctostáphylus Uva-úrsi*. The exquisitely fragrant Mayflower, or Trailing Arbutus, is *Epigéa repens*.

XLII. THE PITCHER PLANT.

Description.—In peat bogs and fresh marshes throughout the country, the Pitcher Plant may be sought. It is everywhere an object of curiosity and wonder. Eight or ten different forms occur, but the flowers in all are exactly similar, except perhaps in color. The species most generally accessible is delineated in Fig. XLII.

Analysis.—GENERIC CHARACTERS.—The habit of these plants is acaulescent, with perennial fibrous roots, leaves

---

Heather-peat. The Heather sprays and blossoms are eaten by grouse and by sheep in a time of scarcity; while the "Heather-bell, with her purple bloom," is a boon to bees.

Fig. XLII.—1, Sarracénia purpùrea.

hollow and containing water. The flowers, one or more, 5-parted, perfect and complete, are large, solitary, mounted and nodding on a naked scape.

The calyx consists of 5 ovate, spreading, colored sepals subtended by 3 bractlets. The corolla is of 5 obovate, incurved petals covering the broad style, and the many hypogynous stamens.

The ovary is glabrous, 5-carpelled, 5-celled; the style short, expanding at the top into a broad umbrella-shaped or *peltate* (i. e., shield-shaped) membrane, bearing the 5 stigmas in the notches of the 5 rhombic lobes. The matured fruit is a capsule with 5 cells opening by 5 valves, having the placentæ in the axis, or inner angle of the cells. The seeds are numerous, anatropous, with a small embryo in much albumen.

THE PITCHER PLANT.

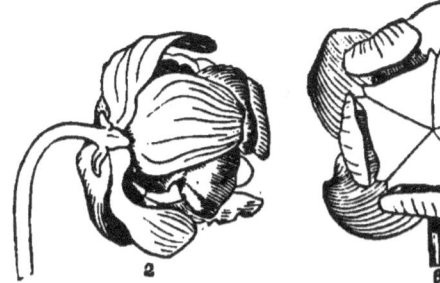

2, a flower seen beneath; 3, a flower seen above; 4, the stamens and pistil; 5, Sarracènia psittacìna; 6, P. Drnmmóndii, leaves only.

Specific Characters.—We have before us (Fig. XLII, 1) the only northern species, distinguished from the others by the leaves alone. These are in the form of a pitcher, 6–9' long, broadest near the middle, as a pitcher should be, ascending, incurved, open, bearing a broad wing along the whole length on the inner side, and at the top an erect cordate, hood-like blade. The hood and much of the tube below is beset within by stiff, sharp, reversed bristles. The capacity is about half a wineglass, and the pitcher is generally filled with water containing drowned insects. The flowers are deep brownish purple, 2–3' broad, on a scape about 1 foot high.

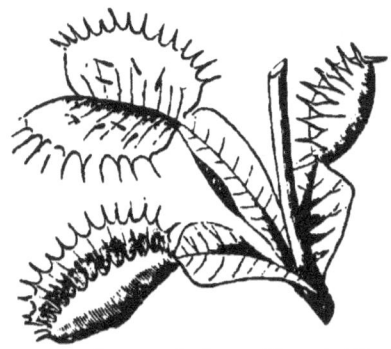

7, Dionæa muscípula, or Venus's Flytrap; Leaves only.

The Name of this genus is *Sarracénia*, conferred in honor of *Dr. Sarrazen* of Quebec, who first sent it to Europe in 1640. The specific name, *purpùrea*, is unfortunate; for its flowers are sometimes yellow, and other species have purple flowers.

The Order SARRACENIACEÆ, the Pitcher Plants, includes 3 genera, viz., Sarracénia, Heliámphora of Guiana, and Darlingtònia of California. All the species (9 or 10) are distinguished for having *ascidia*, that is, leaves which hold water like pitchers, and are probably alike *carnivorous*.

CARNIVOROUS PLANTS.—One of the most wonderful results of recent botanical investigation is to show that certain plants are expressly contrived to entrap and digest insects, and are therefore carnivorous in habit. In Sarracenia the glistening water at the bottom of the pitchers and the sweet secretion on the leaf are fitted to entice inquisitive flies, etc., to alight. Then, the reversed bristles facilitate their entrance, but forbid their return. Finally, the inner surface of the tube secretes a fluid capable of digesting the animal matter and probably also of assimilating it for the growth of the plant.*

XLIII. THE AMERICAN COWSLIP.

Description.—This notable plant adorns our woods and prairies, in May and June, from Pennsylvania westward; but at the approach of the ploughman it flees to the wilderness unless enticed by the gentler arts of the gardener. Its numerous names, as Pride-of-Ohio, Shooting-Star, Dodecatheon, are its titles of nobility; and its aspect, *acaulescent* like Pyrola and Sarracénia, with a radical crown of leaves

---

* Among the other carnivorous plants are Venus's Fly Trap (*Dionœa muscipula*, native of N. Carolina), the Sundew (*Drósera*), and the East Indian Pitcher Plant (*Nepénthes*). In the first named (Fig. XLII, 7) there is a curious trap at the end of the leaf. Along the edges are rows of bristles which have been aptly compared to the eye-lashes. On each side within are three more exceedingly sensitive hairs. If one of these be touched by an insect crawling over the leaf, the two sides will instantly shut upon the hapless prisoner, the fringe on the edge interlacing like the fingers of the two clasped hands. The fluid secreted by the leaf immediately flows out, apparently to aid in the digestion of the animal food thus ingeniously caught. This natural trap may be sprung by dropping into it a piece of meat. In the Darlingtonia there is a bait—an appendage smeared on the inside with honey—hanging at the entrance of the tube, enticing insects to go within.

162   THE AMERICAN COWSLIP.

and a naked columnar scape supporting an involucrate umbel, is the ideal of floral grace and beauty.

The fashion of the Flowers is like that of the garden Cyclamen, otherwise unique, suggesting the thought of a shooting star or a bird on the wing. This effect is due to the white petals so sharply reflexed, while the stamens and style project forward in the form of a parti-colored beak or an arrow-head.*

Analysis.— THE LEAF REGION.—The root is a dense

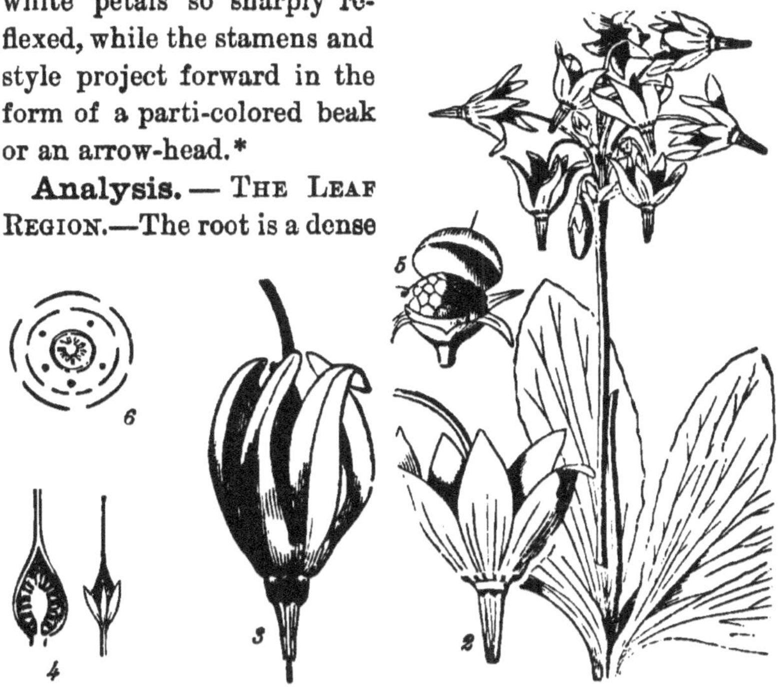

FIG. XLIII.—Dodecátheon Meádia: 2, a flower with pistil undeveloped; 3, a flower, full size, with the pistil; 4, dissection, showing the free-central placenta, &c.; 5, the pyxis of Anagallis; 6, the plan of the flower.

mass of branching fibers issuing from the perennial crown, and striking deep into the soil. The stem (the crown already mentioned) is wholly subterranean, and destitute of

---

* In the Dodecatheon we find two types of flower (dimorphism). In some the pistil is long and flush with the throat of the corolla, and the stamens are fixed half way down the sides; in others, the pistil is short and the stamens are attached to the throat of the corolla. One form has thus the pistil where the other has the stamens. This was long thought to be a mere freak of nature; but it is now known to be another

any definite form. The leaves, sheathing the scape at the base and springing with it from the crown (*radical*), are oblong, obtuse, nearly entire, and smooth. The inflorescence is an umbel.

THE FLOWER REGION.—The Flowers are 5-parted ($\sqrt[5]{}$), complete, perfect, regular, symmetrical, gamopetalous (although the petals are almost separate). What of the calyx ? What of the stamens ? Here is an arrangement like that in Claytònia (p. 41). The 5 stamens stand opposite to (*opposing*) the 5 petals. The slender anthers are coninvent in a slender cone inclosing the thread-like style. The ovary and fruit are superior, 1-celled. The many (∞) seeds are affixed to a central erect column—that is, to a *free central placenta*. Is the ovary simple or compound ? Probably compounded of 5 carpels, since the other organs are in 5s. But the fusion is so intimate as to leave no trace of the seams, lobes, or cells ; nor does the style or stigma give any indications. This is extraordinary. Compare the triple pistil and capsule of Erythrònium (p. 32).

The Name, *Dodecátheon* (*dodeka*, twelve, *theoi*, gods) was conferred by Linnæus as if the flowers (about 12 in number) were so many little divinities—a poetic fancy not unworthy of the great naturalist. *D. Meádia*, the specific name given by Catesby, in honor of Dr. Mead, the discoverer, was originally intended for the genus.

Scientific Terms.—Free central placenta. Opposing stamens.

---

contrivance to secure crossing. An insect lighting upon a short-styled flower would naturally dust its head with pollen from the stamens clustered about the mouth of the tube ; on going to a long-styled flower, its head, covered with pollen, would at once come in contact with the sticky pistil at the opening of the throat ; and *vice versa*, pollen would in the same way be carried from a long-styled flower to fertilize a short-styled one.—It is curious to note also how the flower is, so to speak, " made the most of " in the floral competition for insect services by a simple contrivance. The corolla being deeply cleft and each petal bent backward, brings every part of the surface into conspicuous notice.

XLIV. CHICK WINTERGREEN.

Description.—Dodecátheon is often cultivated in the gardens of New England and New York, but is never native in those States. Its place is there occupied by the pretty

Fig. XLIV.—Trientàlis Americana: 1, a flower; 2, the seeds heaped on the free central placenta.

Chick Wintergreen or Star-flower, growing in the cool, damp woods. No flower in May and June is more lovingly greeted.

Analysis.—The 7-fold division of the floral organs is the most striking feature of this flower. It is seen in the petals, sepals, stamens, and even in the leaves, and probably it exists also in the pistil and fruit. The 7 white, slightly gamopetalous, wide-spread petals, form a wheel- or star-shaped corolla, and the 7 stamens stand *opposing* them. The ovary as well as the style is *one*, and in fruit becomes a 1-celled capsule with about 7 seeds on a free central placenta.

The Name, *Trientális* (*triens*, the third part of a foot) *Americána*, alludes both to the height of the plant and to its native country.*

XLV. THE LOOSESTRIFES.

Description.—There are many kinds of Loosestrife scattered over the country, blooming in June and later. Some choose a gravelly soil, in the borders of woods and thickets. An English writer says, "growing in damp woods, hanging down the sides of mossy slopes, its branches trailing a foot or more long, well clad with roundish, shining, deep-green leaves, and bearing in June and July handsome yellow flowers;" but here they are oftener found in low meadows and miry swamps. One of them, the Moneywort, alluded to above, is cultivated and runs wild in our gardens.

GENERIC CHARACTER.—All the Loosestrifes are perennial herbs, with opposite or whorled (*verticillate*) leaves, and complete, regular, symmetrical, yellow, more or less gamopetalous flowers, generally 5, rarely 6 or 7-parted. The corolla is somewhat wheel-shaped, the stamens as many as, and

---

* We rarely find this plant varying with its flowers 8-parted. In Oregon a variety grows one-third larger, with flowers always 8-parted and rose-colored.

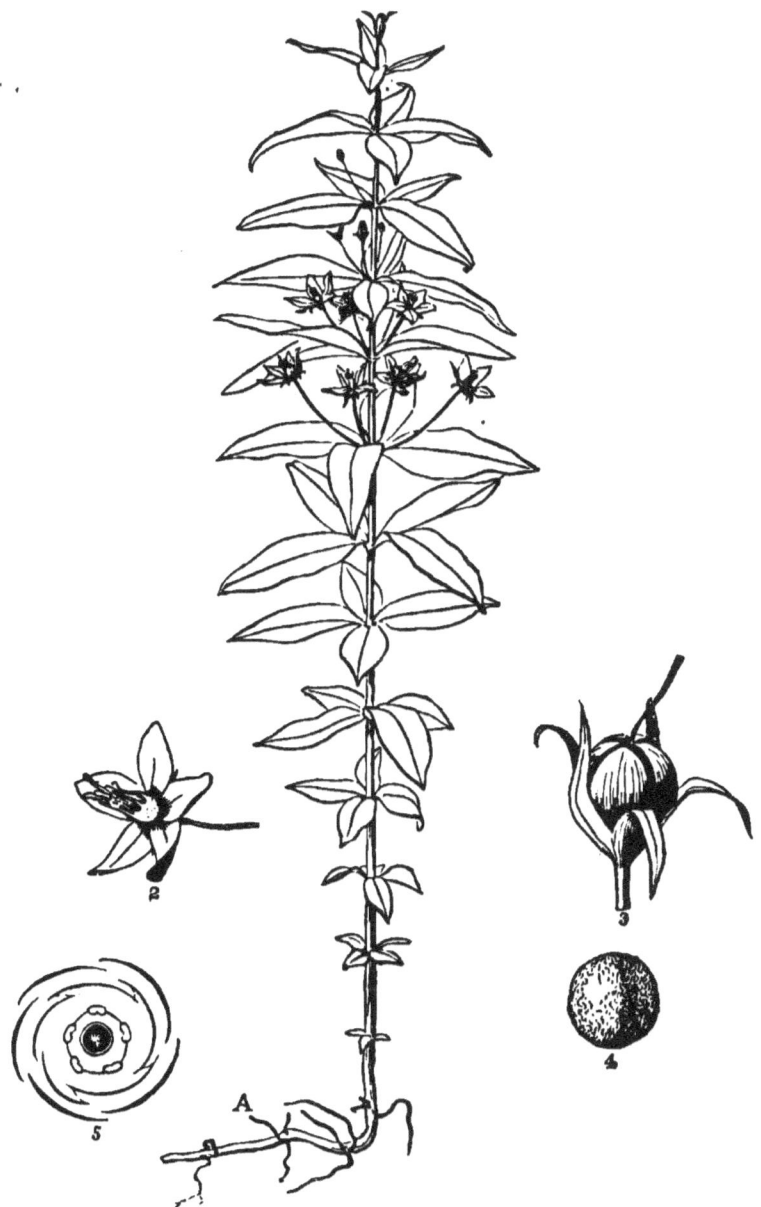

Fig. XLV.—Lysimáchia quadrifólia: 2, a flower; 3, a capsule opening; 4, a seed; 5, plan of the flower.

opposing the petals, generally *monadelphous* (united at the base).* The ovary is evidently compounded of 5 carpels, for in fruit it opens at the top by 5 valves, or 10 half-valves. The seeds are several or many, attached to a free central placenta in the one-celled capsule.

The Name of the genus thus characterized is *Lysimáchia* (Loose-strife or peacemaker), being dedicated to Lysimachus, king of Sicily, "who," says Pliny, "first used it in medicine."

SPECIFIC CHARACTER.—The species whose portrait we give is distinguished from the others as follows: It is an erect, smooth herb, 1–2 feet high, rising from a slender rhizome or creeper, its stem terete, straight, without branches. The leaves, commonly in whorls of 4s, more rarely in 5s, 3s or 6s, are ovate-lanceolate, pointed, very evenly pinniveined, and sessile. The flowers are on *capillary* (very slender, or hair-like) peduncles, one in each axil of the upper leaves, and 5-parted, with the petals longer than the sepals, golden yellow with broken black lines. Stamens of unequal length, evidently *monadelphous*, with no *rudimentary* stamens between. Capsule (in August) globular, 5-valved, few-seeded.

L. quadrifòlia, its specific name, alludes to its whorled leaves. There are other species, as *L. stricta,* with flowers in a terminal raceme, a common plant in grassy meadows. *L. ciliàta,* with the leaves on *ciliate* (hairy-edged) petioles, and larger axillary flowers, with 5 rudiments of stamens, is also common. *L. nummulària,* Moneywort, with trailing

---

* In some species of Loosestrife we shall find certain little points or teeth interposed between the stamens or the petals alternating with both. These are sterile filaments, or rudiments of stamens, and are full of curious instruction. They explain the anomalous position of the stamens in these flowers. With them all the organs alternate. May we not make clear the same anomaly in Dodecatheon and Trientalis by this analogy? We have only to suppose another set of stamens or rudiments *intended*, between the stamens and petals.

stems and rounded leaves, is a handsome foreigner fully naturalized.

In all, we have 10 species. (See *Bot. & Flor.*, p. 212.) The student may record the analysis of any one of them in connection with this lesson.

Classification.— The Order PRIMULACEÆ, the Primworts, represented by the genera Lysimachia, Trientalis, and Dodecatheon, receives its name from the leading genus, Primula, the Primrose.* The following are its attributes:

> Plants low, herbaceous.
> Leaves all radical or mostly opposite.
> Flowers regular, gamopetalous, 5-parted.
> Stamens 5, opposing the 5 corolla lobes or petals.
> Pistils consolidated into a 1-styled, 1-celled ovary.
> Placenta free, central.
> Seeds many or few, with fleshy albumen.

The Primworts include 20 genera and 300 species, of which many are ornamental, especially the Primworts which have long been favorites in the gardens of Europe, and well known in ours. Their numerous varieties are variously called Oxlip, Cowslip, Auricula, Primrose, and Polyanthus.

Cyclamen is native in Syria and Europe. Its round solid bulb (corm) is eaten by swine. Its scapes twist into a coil around the ovary after flowering, and lie close to the ground while the seeds ripen. It is very pretty in pots.

Anagállis, the Poor-man's Weather-glass, is a beautiful trailing weed. It opens its pretty red flowers from 7 to 2 o'clock if the weather be fair, but closes them on a damp or cloudy day. It is noted for its

---

* Name from *primus*, first; for its early bloom. The delicate flowers of some of the 60 species appear when all nature is otherwise inert. They are chiefly natives of Europe, and pre-eminently Alpine. Amid the cold blasts of these dreary regions, where the roots are perhaps bathed in ice-cold water, the little primrose lies secure beneath its fleecy mantle, waiting for a gleam of sunshine only to melt a patch of snow for it to smile forth in all its loveliness of white, yellow, violet, lilac, and sky-blue. A traveller one day passing over the Faulhorn saw a field of snow where a horse had crossed, and the snow disappearing in his tracks, the little circles were brimful of flowers of every hue. Only 2 species are native in the United States, and these so rare that the collector looks for them as for treasure.

LYSIMACHIA.

seed-vessel—a *pyxis* (pp. 43, 162) opening like a snuff-box. Being found throughout the United States, though not abundant, we give its *record* as a model for the Primworts:

ORGAN.	Life, Habit, Number, Place, Dehiscence, Kind, Construction, Form, Placentation, Size, Qualities, Appendages.
Plant, L.H.S.Q.	☉, *low, diffusely spreading, 6–15', smooth.*
Root, L.K.	☉, *axial, branching and fibrous.*
Stem, L.H.K.F.	☉, *procumbent, branching, herbaceous, quadrangular.*
Leaves, L.P.C.F.S.Q.	*Opposite, sessile, palmi-veined, ovate, entire, smooth, 8–12".*
Inflorescence, P.K.A.	*Axillary, solitary, peduncles longer than the leaves.*
Flower, N.C.	*Many, perfect, complete, regular, 5-parted.*
Calyx, F.Q.	*Rotate, spreading, green, smooth.*
Sepals, L.N.P.F.	*Persistent, 5, united at base, lanceolate-linear.*
Corolla, F.Q.	*Rotate, spreading, red or blue, or white, minutely fringed.*
Petals, L.N.P.F.	*Deciduous, 5, spreading in sunshine, obovate, united at base.*
Stamens, N.P.C.	*5, hypogynous, opposing the petals, filaments bearded.*
Anther, D.C.F.	*2-celled, opening lengthwise, oval.*
Style, N.C.F.	*Single, very short.*
Stigma, N.F.	*Single, capitate.*
Ovary, C.F.Pn.	*Compound, indivisible, 1-celled, with free central placenta.*
Fruit, N.D.K.F.Q.	*Single, a pyxis, opening by a lid, circumscissile.*
Seed, N.C.F.Q.A.	*Many, albuminous, angular, rough.*

LOCALITY.—*East New York, L. I.* (Date), *June 20, 1878.*

CLASSIFICATION.—**PHENOGAMIA ; GAMOPETALOUS EXOGENS.**

ORDER.—PRIMULACEÆ—THE PRIMWORTS.

NAME.—Latin, **Anagallis arvensis.**

—English, *Poor-man's Weather-glass.*

REMARKS.—*The flowers open from 7 to 2 o'clock if the weather be fair, but close on a cloudy or rainy day.*

XLVI. THE SPEEDWELLS.

Description.—Along the borders of the woods, in the hilly pastures, the open fields, and even in the waste corners of the garden, we often meet the smaller Speedwells. Their tiny flowers greet us with a clear, honest welcome in the dewy mornings of May and June.* We shall know them by their 4-parted corolla, 2 stamens, double ovary, and by the following more definite

GENERIC CHARACTERS.—The Speedwells are small or large herbs, with opposite or whorled leaves, and small blue, white, or reddish flowers in the axils of the leaves, or in terminal racemes. A 4-parted, green calyx supports a gamopetalous, colored corolla which is 4-parted, slightly irregular, with the upper lobe somewhat enlarged. There are but two stamens, placed one on each side of the upper lobe and *exserted* (projecting). The pistil is evidently double, for though only 1 style and 1 stigma appear, yet the ovary is 2-lobed and 2-celled, and the fruit a 2-celled, ∞-seeded, flattened capsule.

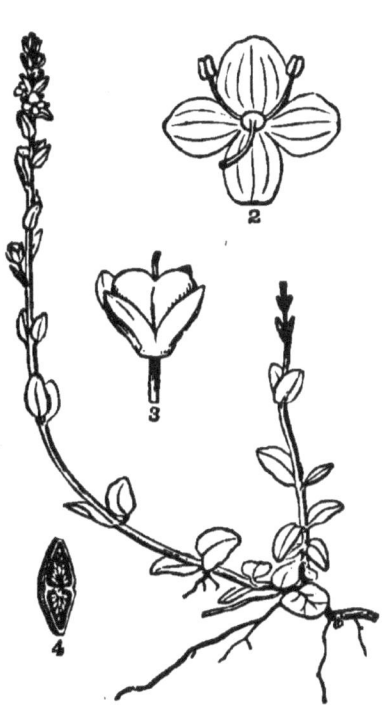

FIG. XLVI.—Verónica serpyllifòlia: 2, a flower; 3, a capsule with the persistent calyx; 4, section of the 2-celled capsule.

---

* Among the Germans, the Speedwell is known as the Flower of Truth, and the plant is taken as the emblem of friendship. The popular name, "Speedwell," is a parting salutation, equivalent to "Farewell!" "Good-bye!" It comes (says Prior) from the *evanescent* corollas, which fall off and fly away as soon as the plant is gathered.

SPECIFIC CHARACTERS.—The species represented in Fig. XLVI, abounds along road-sides, in hilly pastures, and old grass-plots. The stem ascends 2'-6' from a reclining, branching base, with leaves opposite, slightly crenate, the lower petiolate, roundish, the upper becoming oval, oblong, and bract-like. The flowers form a loose, terminal raceme, in which each is located in the axil of a bract. The corolla is *rotate* (wheel-shaped), white, penciled with blue lines, scarcely more than 1" in breadth; and the pods roundish, *retusely* notched.

The Name is *Verónica serpyllifòlia* (V. the Thyme-leaved). The genus is said to have been dedicated to St. Veronica* (*vera*, true, *icon*, image). It is extensive, embracing 150 species growing in the cooler parts of the earth. The student will meet them everywhere, and may as profitably analyze the following as the foregoing:

V. peregrìna (the Foreigner; so it acts, but it is native), known as Purslane Speedwell, ① or ②, 5–9' high, smoothish, with oblong, toothed leaves, whitish wheel-shaped flowers, and notched pods.

V. arvénsis (the Field or Corn Speedwell) ①, hairy, with roundish and ovate, crenate leaves, pale blue flowers, and obcordate pods. With the first, it abounds in cultivated grounds.

V. officinàlis (the Official S.) a ♃ in woods and pastures, ascends 6–12' from its decumbent base, with oval, obtuse, *serrulate* (finely serrate) leaves, and the flowers in a terminal raceme.

V. Scutallària will be recognized by its long lance-linear leaves and axillary racemes with filiform stalks, growing in swampy places.

---

* In ancient tradition, St. Veronica is represented as the daughter of Salome. When she witnessed the procession to Calvary with Christ bearing his cross, she wiped the drops of agony from his brow; and thenceforth the image of the Saviour was miraculously imprinted on the napkin.

THE SPEEDWELLS.

ORGAN.	Life, Habit, Number, Place, Dehiscence, Kind, Construction, Form, Placentation, Size, Qualities, Appendages.
Plant, L.H.S.Q.	♃, herb, 6–12′, ascending, pubescent.
Root, L.K.	♃, fibers clustered at the nodes of the creepers.
Stem, L.H.K.F.	Herbaceous, decumbent at base, caulis aerial.
Leaves, L.P.C.F.S.Q.	Cauline, opp. pinni-veined, pet., oval, obtuse, serrulate.
Inflorescence, P.K.A.	Terminal raceme pedunculate.
Flower, N.C.K.	4-parted, ☿, fertile, irregular, gamopetalous, 1¼″ diam.
Calyx, F.Q.	Rotate, 4-cleft, green, hairy.
Sepals, L.N.P.F.	Persistent, 4, valvate, spreading, oblong.
Corolla, F.Q.	Caducous, rotate, 4-lobed, white, with blue lines.
Petals, L.N.P.F.	Caducous, 4, imbricate, lowest one smallest.
Stamens, N.P.C.	2, exserted, epipetalous (on the petals).
Anther, D.C.F.	Cells 2, confluent at apex.
Style, N.C.F.	One, double, thread-form.
Stigma, N.F.	One, double, 2-lobed.
Ovary, C.F.Pn.	Compound, 2-celled, superior, obcordate.
Fruit, N.D.K.F.Q.	2-carpelled, 2-celled, capsule, valvate.
Seed, N.C.F.Q.A.	Few, albuminous, ovate, smooth.

LOCALITY.—*Rocky woods, West Farms, N. Y.* (Date) *June 12, 1878.*

CLASSIFICATION.—GAMOPETALOUS EXOGENS.

    ORDER.—SCROPHULARIACEÆ, or FIGWORTS.

        NAME.—Latin, **Veronica officinalis**.

           —English, *Officinal Speedwell.*

REMARKS.—*The plant is bitter and astringent, used for tea in Europe, hence officinal.*

XLVII. THE TOAD FLAX.

Description.—The wayside and the borders of fields are often ablaze with the Summer robes of the Toad Flax. As it is wont to grow in large, dense patches, the collector signals its flame-colored flowers at a long distance.

Analysis.—THE LEAF REGION.—For the Record, the student will determine the life and kind of the root, the arrangement, construction, form, and quality of surface of the leaves, and the place, kind, etc., of the inflorescence. No new nor striking feature is noticeable in the parts constituting this portion of the plant.

THE FLOWER REGION.—The flowers always attract the attention of the curious by not only their brilliant, showy hues, but also their singular structure, both of which justify the popular names,—Snapdragon, Butter-and-Eggs, etc.

The small green calyx indicates a *pentamerous* (5-parted) tendency in the flowers, and the oddly-shaped corolla gives a faint echo of the same by the 5 unequal lobes of its border. An inflated tube ends in a mouth and lips above, and in a tail (spur) behind. The lower (outer) lip is 3-cleft, the upper 2-lobed. The throat is closed by the prominent orange-colored palate. If lateral pressure is applied, it gapes, and closes again with a snap. In technical language, the corolla is *bi-labiate* (*bis*, two, *labia*, lips), or simply *labiate*. For its *closed* throat it is *personate* (*persona*, a mask), and for its tail, *spurred.**

---

* The spur is the nectary and the entrance to it is generally closed by hairs. The nectar, therefore, can be reached only by insects having a long proboscis. Thus again is the student reminded of the mutual adaptation of flowers and insects. Here, as in the Evening Primrose and Honeysuckle, is a rich store of nectar; but it is deeply hidden in the long spur or tube, while the flower gives off its strongest fragrance at night. Now it is at night that the Sphynx Moths fly abroad. They have long tapering wings that enable them to poise for a long time in one position. At-

The *Stamens* (2) are *didynamous* (*dis,* two, *dynamis,* power), two of them being longer than the other two. The law of symmetry would require a 5th stamen, as it does a 3d and 4th in Veronica.*

The *Ovary* is in the midst of all (2), surmounted by a slender style, and maturing to an oblong capsule (3) of 2 cells (4). The many seeds are *wing-margined* (5), escaping finally by chinks opening between the thin valves.

In the figure (6) is represented a seed dissected, showing a straight 2-lobed embryo in copious albumen.

The Name is *Linària vulgàris*. Linaria alludes to its general likeness to the Flax (*Linum,* whence the word *linen*); vulgaris is given because it is common—too common indeed, throughout Europe, Asia, and America, for it often grows to an army of intrusive weeds difficult to extirpate by reason of its long creeping roots.† Another species, *L. Cymbalària*—the pretty Ivy-leaved Toad Flax, is often seen in the greenhouse and parlor.‡

Classification.—These genera, Linaria and Veronica, represent the great Order SCROPULARIACEÆ or Figworts.

---

tracted by the light color and the powerful odor, they hover over the plant, while they thrust their long sucking trunk into tube after tube as they flit about, apparently robbing the plant of its honey, but really serving the very end of Nature as pollen-bearers.

* In Pentstemon, a nearly related genus, the 5th stamen appears as a filament without an anther, and in Mullein, of the same order, the 5th stamen is complete.

† Mr. Watson, in his Annals of Philadelphia, says that it was introduced from Wales, as a garden flower, by a Mr. Ranstead, a Welsh resident. Hence one of its popular names, *Ranstead*. This plant may remind us that not everything in Nature was designed for the use of man alone. Flowers grew, blossomed, and bore fruit in the geologic ages, before man was created. The colors, odors, and forms of flowers are made to subserve ends of their own. We may delight in these beautiful floral contrivances even without knowing their design in the economy of the plant; but greater should be our admiration when we discover that by a wise frugality of means the beautiful is also the useful and the necessary!

‡ "The capsules of our Ivy-leaved Toad-flax (*Linària Cymbalària*) before ripening turn round toward the wall on which the plant so often grows and creeps, and place themselves in a crevice or hole, so as to shed the seeds, when ripened, in a place where they will thrive, instead of scattering them on the ground where they would be wasted."—*Pratt's Flowering Plants of Great Britain.*

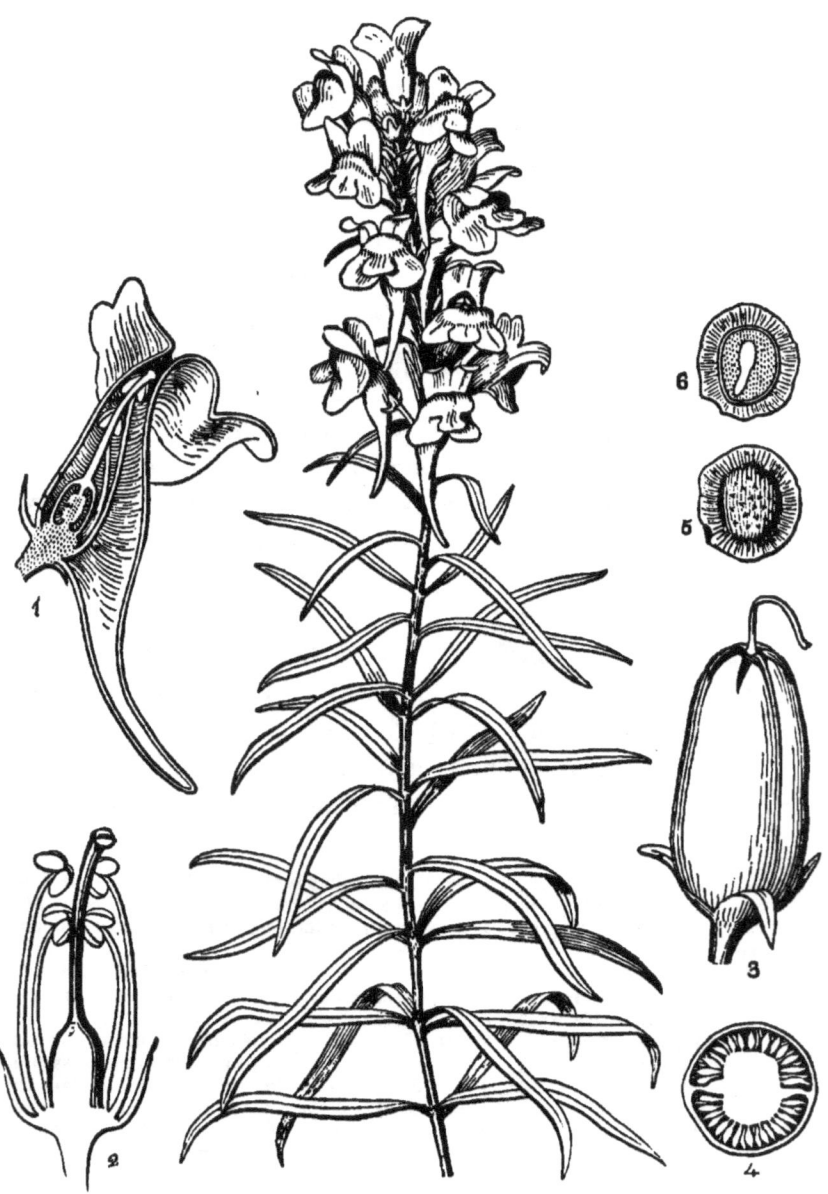

Fig. XLVII.—Linària vulgàris: 1, section of the flower, showing the arrangement of the organs; 2, the stamens and pistil; 3, the capsule; 4, its cross-section; 5, a seed; 6, a seed dissected.

Here also belong the Mulleins, Foxgloves, Gerardias, Pentstemons, and all other plants which possess the following seven characteristics.

> Flowers irregular, without fragrance.
> Calyx free, persistent.
> Corolla gamopetalous, imbricated in æstivation.
> Stamens 2 or 4, rarely 5, inserted on the corolla.
> Ovary free, double, with 1 style and a 2-lobed stigma.
> Fruit a 2-celled capsule with axial placentæ.
> Seeds many, anatropous, albuminous.

The Figworts include 157 genera, 1800 species, abounding in all climes and countries. Among them are some medicinal and poisonous plants, as *Digitális* (Foxglove),* and many cultivated for their handsome flowers, as *Calceolária* (Ladies-slipper), *Antirrhinum* (Snapdragon), *Pentstemons, Maurandias, Russellias.*

The Officinal Speedwell (*V. officinális*) was formerly used as tea in Europe, but there is no Figwort considered truly nutritious or useful for food.†

Scientific Terms.—Bilabiate. Didynamous. Labiate. Pentamerous. Personate. Spurred. Wing-margined.

XLVIII. THE GROUND IVY.

Description.—This interesting plant, like the Dandelion and other naturalized foreigners,‡ selects its home in

---

* This term is generally supposed to be a corruption of Folk's or Fairies' Glove, these imaginary beings having formerly been known as the "good folk." There are many superstitions attached to the plant and it is still thought by the ignorant to be a favorite lurking place of the fairies. In South Wales the children are wont to hold one end of the Digitalis bell and strike the other with the hand to hear the fairy thunder with which the indignant little sprite is supposed to make its escape from its injured retreat. According to some legends, the fairies lend the blossoms to the fox on his marauding expeditions, to soften his already velvet tread.

† During the famous siege of Rochelle by Richelieu, in 1628, the garrison for a time lived entirely on the root of a kind of Figwort, probably the Scrophularia aquatica. From this circumstance the plant is known in France as Herbe de Siege.

‡ Let us carefully distinguish between our *native* and *naturalized* plants. The former are characteristic of the country, and have flourished in its wilds, independent of man, for unknown ages. Such are Dodecatheon, the American Elm, &c. Naturalized plants once introduced from other lands, whether by accident or design, find

cultivated soil. It prefers shady or stony places in parks, fence-rows and rubbish, and grows with vigor, blooming from May to August. It is a smooth perennial (⑭), here prostrate on the ground only, though in Europe it is often seen, with Moss and the True Ivy mantling the garden wall and ancient ruin.*

Analysis.—What of the *Roots?* The slender square stems creep extensively, forming loose mats, and putting forth at each node a pair of leaves and a tuft of fibrous roots.

The *Leaves* are all of one pattern, opposite, long-petioled, palmi-veined, round-*reniform* (kidney-shaped), *crenate*, i. e., with rounded teeth, on the margin.

The large blue *Flowers* appear in loose axillary clusters. The calyx is tubular, slightly curved, 15-veined, obliquely 5-toothed; corolla a thrice-longer tube, 1' long, bilabiate, upper lip 2-lobed, lower 3-lobed, with the middle lobe largest. Looking within the corolla we find 4 didynamous stamens, as in Linaria, 1 less than symmetry requires. They stand in pairs tending toward the upper side, the inner pair longer than the outer. The anther comprises 2 separate lobes diverging at right angles, so that each pair in contact forms a perfect cross. There is one slender style with a 4-parted ovary.

---

here a soil and climate congenial to their nature, and grow spontaneously, as well as, or even better, than in their own country. Such are the Dandelion, Mullein, Shepherd's Purse, Apple-tree. They generally betray their origin by their habits, planting themselves in gardens, fields, highways, wherever the soil has been stirred by the plough, or trampled by the foot of man. The Indians call our Common Plantain "the White Man's Plant," and say it springs up in his trail, wherever he plants his foot.

* In 1850, a deputation waited upon the Chancellor of the Exchequer in England respecting an abolition of duties on window-glass. To enforce their views as to the deleterious effects of unlighted dwellings, they exhibited a Ground Ivy plant, which had grown for some years in a Wardian case on the top of a model of an abbey. The branches which were turned toward the light were laden with leaves, flowers and fruit; while the stems which had trailed down between the model and the window, and so lost the light, had no blossoms or fruit, and their leaves were scarce one-tenth as large as the others. Every condition of growth, save that of sunlight, was necessarily the same for all the branches of the plant, and the dwarfed, starved state of one portion arose solely from that single deprivation.

The *Fruit* consists apparently of 4 reddish oval seeds contained, until ripe, in the persistent calyx. But the *seeds* must not be confounded with the fruits which contain them. There are 4 achenia or *nutlets*, each containing 1 seed.

The Catmint, blooming early in July, will also fall in

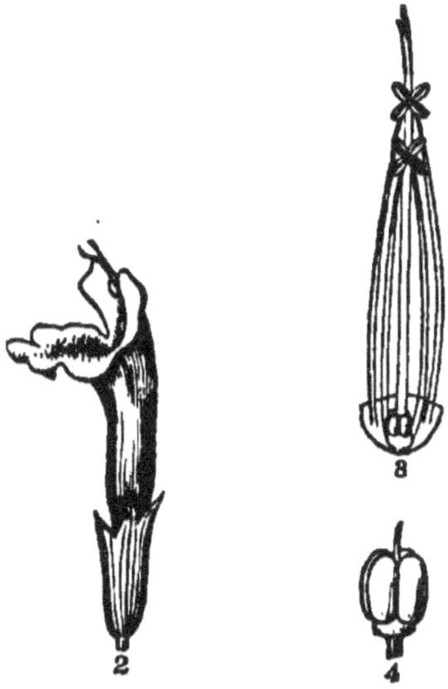

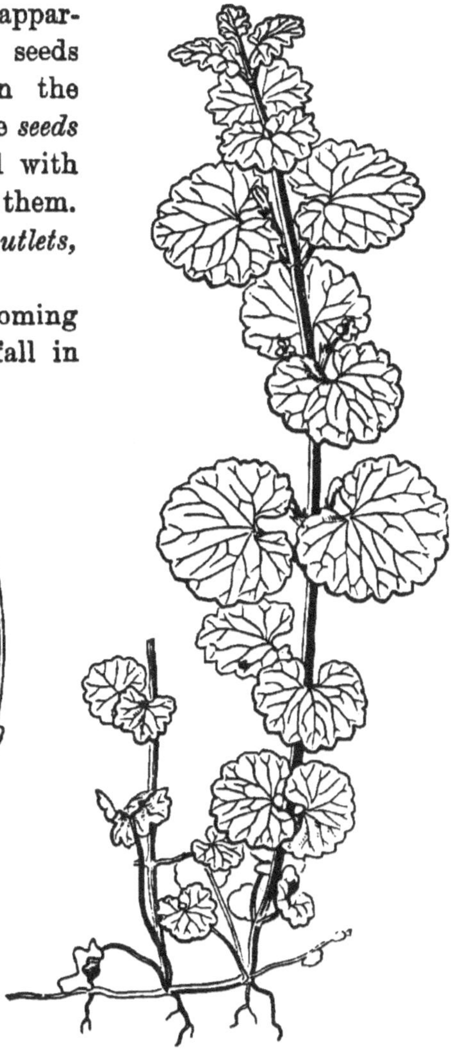

Fig. XLVIII.—Népeta Glechòma: 2, a 'flower; 3, the stamens and pistil; 4, the fruit—four achenia.

the way of every collector. It is another foreigner, perfectly naturalized, springing up in waste corners around our country dwellings. Let the student compare the Ground Ivy and the Catmint,* and carefully note the *resemblances* and the *differences*. The former will make up the *generic*, the latter the *specific* characters—thus :

RESEMBLANCES (generic). In both, the stem is square; leaves opposite; calyx tubular, 15-veined; corolla bilabiate; throat not hairy, upper lip 2-lobed, lower 3-cleft, middle lobe largest; stamens ascending; anthers approximating by pairs, their 2 cells separate and diverging; style bifid, fruit 4 achenia.

DIFFERENCES (specific). The Catmint is clothed with a whitish pubescence; the stem is erect; the leaves short-petioled, ovate, cordate, acute, crenate-serrate, the upper reduced to bracts; the flowers in dense axillary clusters (*verticils*); corolla *not* twice longer than the calyx, white.

The Name of the genus thus characterized is *Népeta* (*Nepet*, a town in Tuscany). Ground Ivy is *N. Glechòma* (ancient Greek for Thyme). Catmint is *N. Catària*, a play on Puss's name, whose fondness for the herb is thereby commemorated.

XLIX. BLUE CURLS.

Description.—This plant is perhaps better known by the name Self-heal.† It is a native of low grounds both in fields and forests, flowering from May to August according to climate. Its squarish, blue-flowered cluster is a familiar object in the rural scenes of our boyhood. In New England its growth is stinted to a few inches in stature, but in the rich bottoms of the West it attains to several feet, its flowers being proportionately larger.

---

* Dried specimens of Catmint may be used for comparison, when fresh ones in flower cannot be found.

† The popular name, Self-heal, intimates that with it one may cure himself, or as expressed in the French proverb quoted by Ruellius (*De Natura Stirpium*), "No one needs a surgeon who has Prunella."

BLUE CURLS.

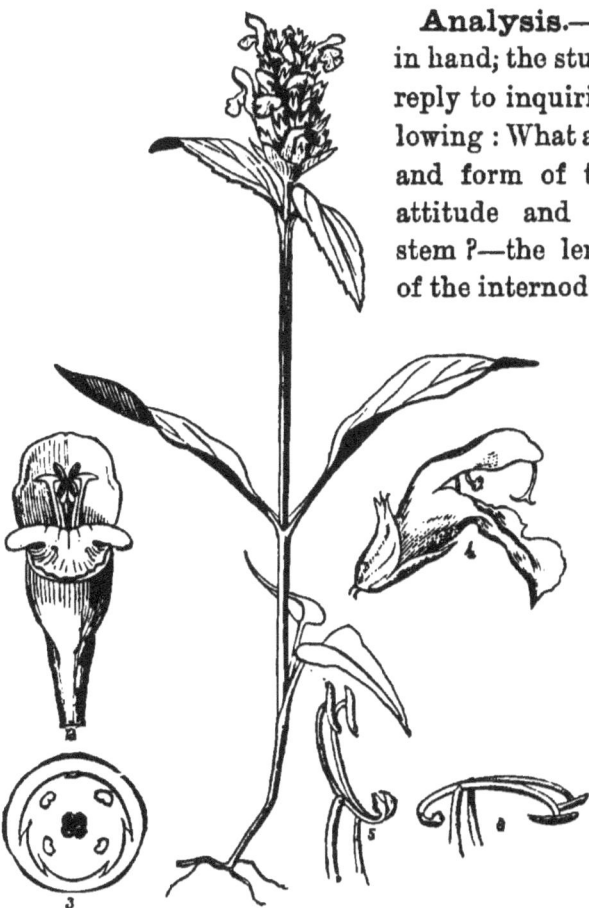

FIG. XLIX.—Brunélla vulgàris : 2, a flower without the calyx, showing the spurred filaments ; 3, plan of the flower ; 4, a flower of Sage (Sálvia) ; 5, the 2 stamens in their natural position ; 6, after being tilted over by a bee.*

Analysis.—With the plant in hand; the student will easily reply to inquiries like the following : What are the duration and form of the root?—the attitude and height of the stem ?—the length and form of the internodes ?—branches? What is the clothing of the plant ? What is the arrangement of the leaves? — length of petiole ?— form of the blade?—margin ?—apex ? base?—venation ?

---

* In the common garden Sage (Salvia officinàlis), there is a curious device for securing cross-fertilization. There are but 2 stamens; the 2 cells of each anther, instead of being close together as is usual, are widely separated by a long *connective* (5, 6). The lower cell contains very little if any pollen, while the upper is full. The connective is fixed to the filament by a pivot, and naturally stands in position as seen in 5. Meanwhile the stigma is yet immature and high up in the arch, when a bee seeking nectar alights on the door-step—the lower lip—and entering the tube pushes against the lower anther cells, tilts the connective as seen in 6, bringing the upper cells down on his back. The next flower he visits has perhaps its stigma mature and situated as seen in 4, occupying the same place which was before occupied by the tilted anthers, which have now withered away. The learner may observe these phenomena for himself.

What is the form of the bracts?—color of the flower?—form of corolla?—upper lip?—lower lip?—number of the stamens?—construction?—which the longer pair?—appendage of the filament?—5th stamen? How many styles are there?—what are the kind and form of the fruit?

Observe that the leaves are rather obtuse than acute; that the broad bracts are palmi-veined and tipped with a cusp (*cuspidate*), and the hairs are jointed.

Inflorescence.—The flowers occur in 3s, each triplet occupying the axil of a bract, and the middle flower opening first according to the centrifugal mode. Such a cluster is a *cyme*. Many such, with their bracts, are closely imbricated, forming a dense terminal, 4-sided *spike* (for the flowers are sessile).

The *Flowers*. The calyx is colored, bell-form (*campanulate*), 2-lipped, the upper lip *truncate* (square-cut), with 3 small teeth, the lower lip 2-cleft. In the corolla, observe the *vaulted* or concave upper lip, covering the stamens and style, the lower, 3-lobed, dependent lip, and the ring obstructing the tube within near the base.* The longer pair of stamens is the lower (outer), and a *spur* or *tooth* appears on each filament above near the 2-parted anther. Four egg-shaped achenia are at length found in the bottom of the calyx, as in Nepeta.

The Name in science is *Brunélla vulgàris;* Brunella, from the German *braeun*, the quinsy; this plant being a reputed remedy for this disease; *vulgaris*, common; since it grows in nearly every country on the globe.

Classification.—Nepeta and Brunella are now seen to be closely related. Features which they possess in common

---

* In the labiate flowers it is noticeable how the lower lip is arranged for the convenience of insects alighting, and how all the flowers are so grouped as to give this *doorstep* the utmost prominence.

characterize the vast and important order of Labiate Plants, or LABIATÆ.

> Herbs aromatic.
> Stems quadrangular.
> Leaves opposite, exstipulate.
> Corolla bilabiate more or less.
> Stamens didynamous or diandrous.
> Ovary deeply 4-parted.
> Fruit 4 nutlets or achenia.

The Labiate Plants include 125 genera, 2550 species. Among them are the Mints (*Mentha*)—Peppermint, Spearmint, etc.; also Hoarhound and Hyssop, Balm and Lavender, Sage and Pennyroyal. Their richly aromatic oils are stimulant; their extracts febrifugal, None are poisonous. The Oil of Peppermint, the best known among essential oils, is obtained by distillation from *Mentha piperita*. Oil of Lavender is distilled from *Lavándula vera*, and Oil of Spike from *L. Spica*. The former is used in perfumery, the latter in delicate varnishes, etc.

L. MORNING GLORY.

> "*O bells of triumph! delicate trumpets, thrown*
> *Heavenward and earthward, turned East, West, North, South,*
> *In lavish beauty! Who through you hath blown*
> *The sweet cheer of the Morning?*"   CELIA THAXTER.

Description.—This glorious plant is a native of Tropical America and now universally cultivated. It is also nearly naturalized with us, growing spontaneously as a weed from seeds shed in cultivated grounds. It is strictly annual. In a single season it accomplishes its wonderful growth, transfers its own vitality to a thousand seeds, and dies.

The Flowers are ephemeral (*epi* for, *hemera* a day). Beginning to open soon after midnight, they greet the Sun at his rising, arrayed in all their glory, and before he reaches the meridian, fold their robes and perish. But their work

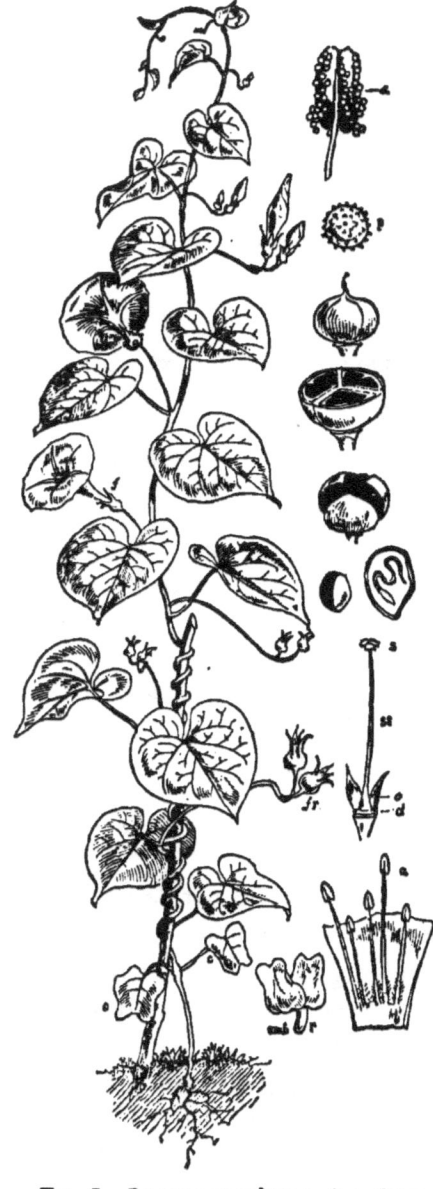

is done, and their successors, already in bud, will renew the gorgeous display the following morning.

Analysis. — Observation and study are wont to begin with the *Flower*, and for once we will reverse our usual order. The *calyx*, the outer envelope, green, persistent, is composed of 5 long-pointed sepals combined at the base into a cup. The inner envelope, the fugacious *corolla*, between trumpet and bell-form, of delicate texture and intense colors, is composed of 5 broad petals united along the *plicate* (folded) edges, quite to the expanding entire border. In the bud, the folds are *contorted* (twisted) *with the sun*, i.e., from left to right—a kind of æstivation called *supervolute*.

The 5 *stamens* adhere to the lower part of the corolla tube, opposite to its

FIG. L.—Ipomœa purpùrea: *f*, a flower; *sta*, the stamens; *st*, the style; *s*, stigma; *o*, ovary; *d*, disk; *a*, anther and pollen; *p*, a pollen grain; *fr*, capsule; *c*, capsule dissected; *o*, capsule opening; *sd*, seed; *z*, embryo; *emb*, the embryo growing; *r*, radicle; *c*, cotyledons.

folds, and fall with it. The 2-celled anther is *adnate* (fixed laterally) to the filament above. Opening lengthwise, the cells disclose innumerable round, white grains of pollen, which, under the microscope, appear beset all over with blunt points or tubercles.

The *style*—the central organ — smooth, slender, supports the 3-lobed stigma at the top, and stands upon the free ovary. The *disk,** a fleshy ring, begirts the ovary at its base; hence it is annular and hypogynous. The contents of the ovary will be understood by viewing its various *sections* (cuttings) under a lens, when 3 cells, each with 2 ovules (young seeds) will be seen. Let the student observe the attitude of the ovules, the place and the organ whence they arise. Their destiny we well know. They will become the seeds in the ripening fruit, and from them new plants will arise the following year.

The *pollen.* Watch the expanded flower at sunrise. The anther cells are also open, and the pollen is set free, to fall, to fly with the wind, or be rudely brushed away by the humble-bee as he plunges into flower after flower in search of the nectar secreted in its depths. Thus a thousand grains may be lost, but some few are almost certain to be lodged on the stigma standing in the midst. On this event depends

---

* A disk is an outgrowth of the torus under or around the ovary. It may be annular, or cup-form, according to the degree of its development. When it does not adhere to the ovary or calyx, it is said to be free and *hypogynous ;* when it adheres to the base of the calyx it is *perigynous.* Sometimes it adheres to both the calyx and the ovary, gluing them together, and even enlarging on the top of the ovary, as in the Umbelliferæ ; then it is *epigynous.*

the life and growth, i. e., the *fertilization* of the seed. If the stigma be covered or destroyed so as to prevent the action of the pollen, no seed will be perfected in the ovary and no fruit produced. Or if the stigma remain good and yet no pollen be lodged upon it, the fruit is equally sure to fail. Therefore the nectar secreted in the nectaries of the flower, and the insect that comes to gather it while unconsciously scattering the pollen, are both necessary links in the Creator's plan. Thus the flower is not merely a thing of beauty. It is an apparatus for a specific work in which each organ performs a definite part. That work is the production of living seed for the perpetuation of its kind upon the earth.

The *Fruit*. After the corolla with the stamens has fallen, the calyx folds itself closely on the ovary and covers it while both continue to grow. At maturity the calyx again spreads and discloses a dry, round pod—a capsule, of curious and beautiful structure. It appears 3-carpelled and 3-celled, as predicted by the 3-lobed stigma. The 3 valves separate at the lines of their juncture with the partitions (a *septifragal* dehiscence), leaving the latter persistent, entire.

Seeds. We find in each cell 2 seeds, the perfected work of the flower. Their structure may be observed by tearing one open just before it becomes hardened, or by sections cut in various directions. Here is a pair of oddly shaped, greenish leaves joined to a short stem, folded and packed with a gelatinous substance. It is the *embryo*, or young plant, and its nourishing *albumen*.* How does this seed differ from

---

* The question of a seed's vitality is interesting, at least to the gardener. He accepts all kinds as good for a year, and, as a rule, rejects such as are known to be older. There are, however, many kinds of seeds which are long-lived. The seeds of Maize and Rye have been known to grow after 30 or 40 years old; Kidney Beans when 100, and the Raspberry (according to Lindley) after 1700 years. It is often observed that when, from deep excavations, earths are first brought to the surface, they are soon covered with strange plants, probably from seeds long buried. After the "Great Fire in London," the Hedge Mustard (*Sisymbrium*), previously unknown in that locality, sprang up thickly amid the blackened ruins.

that of the Apple (p. 111), or the Pea (p. 118) ? It has albumen separate from the embryo, while in the Apple seed and Pea there is no separate albumen, but the nutritive matter is stored up in the massive cotyledons. Hence that important distinction in seeds—the *albuminous*, and *exalbuminous*.

GERMINATION.—In the Spring months you will find the seeds of the Morning Glory germinating in almost every garden. Our cuts show them in various stages. The seed has absorbed water from the soil. The embryo and albumen are softened; the latter is sweetened, and so imbibed by the growing radicle which soon protrudes and turns downward. The cotyledons enlarge, burst the seed-coats, and spread skyward as a pair of leaves (*c, c*). In the axis between them a bud appears, grows, and in a few days its outer scales begin to unfold in succession as a 3d, 4th, and 5th leaf, while the axis extends into internodes between. Thus leaf after leaf, in the order of a spiral line, is unfolded, while the axis with its ever-growing bud at the summit still mounts higher.

BRANCHES.—By this process the one terminal bud is developed without limit into a plant with a simple stem. At length other buds appear, one in the axil of each leaf. From these axillary buds come the branches and flower-stalks.*

A CLIMBER.—The weakness of the Morning Glory vine is compensated by its wonderful instinct. Unable of itself to stand upright, it creeps toward the nearest support and ascends by twining around it spirally. The direction of its

---

* Carefully examined, the seed, or starting-point in the life of the plant, is composed of a leaf, or leaves, closely packed, and altered in tissue and contents so as to suit its new requirements. This is shown in the germination of a Bean or Morning Glory, where the two seed-lobes (cotyledons) arise with the stem as leaves nourishing the young plant. In the Pea they remain stationary at the base of the stem, yielding their nourishment but never expanding. The bud, which, like the seed, is an epitome of the plant, is also composed of leaf-rudiments closely folded, and protected from Winter frosts by thick leathery scales, and evolving in Spring the stem, leaves, and fruit—in short, every structure which comes from the seed.

turning is always *against the sun*—from right to left, contrary to the twisting of its corolla buds.*

The **Root** has no such aspiration. Growing downward from the first moment of its breaking through the seed-coats, it persistently avoids the air and light, seeking the dark, damp depths of the soil. Its innumerable fibers are so many mouths absorbing water and earthy matters, which ascend and mix with the air and gases absorbed by the leaves. Chemical action is induced by the rays of the Sun, transforming all into nourishing sap for the life and growth of every part of the plant.

The Name.—By the latest authorities (*Bentham & Hooker's Genera*), the Morning Glory is called *Ipomœa purpùrea*. But it has many synonyms. In 1750, Linnæus first named it *Convólvulus purpùreus*. In 1790, Lamark transferred it to the genus Ipomœa. In 1840, Choisy separated it, together with all other 3-carpelled species, from Ipomœa to his new genus, Phárbitis. *Ipomœa* is from *ips*, Greek for Bindweed, *omœos*, like.

Classification. — The order CONVOLVULACEÆ — the Bindweeds—represented by the Morning Glory, is limited as follows :

> Herbs trailing or climbing, with alternate leaves.
> Flowers regular, 5-parted, perfect.
> Calyx of 5 sepals imbricated in æstivation.
> Corolla of 5 united petals, supervolute in æstivation.
> Stamens 5, unequal, adhering to the corolla tube.
> Ovary and capsule 2 or 3-carpelled, 2–4-celled.
> Seeds with large embryo and thin albumen.

The Bindweeds number 32 genera and 800 species, chiefly inhabiting the warm regions of the globe.

The Sweet Potato is *Batátas édulis*, a vine resembling the Morning

---

* It seems to be a common law among twining vines that each species should twine invariably in one direction—some (as the Hop) always with the Sun, others (as Morning Glory) against the Sun.

Glory, said to be a native of India. It is cultivated by cuttings, and seldom flowers. The potatoes are tubers growing from the stem as short underground branches. They serve the plant as reservoirs of surplus starch and sugar for its use in early Spring.

Jalap, a well-known drug, is the root of *Ipomœa purga* of Mexico.

Scammony is the root of *Convólvulus Scammònium* of Syria.

LI. THE ROCK MAPLE.

Description.—This valuable tree, known as the Rock Maple or Sugar Maple, grows in forests, openings, or fields, from Canada to the mountains of Georgia, and from Nova Scotia to the Rocky Mountains. It is most abundant in the New England States, where it is an embellishment

Fig. LI.—Acer saccharinum — the Sugar Maple. Sugar-making in New Hampshire.

in almost every landscape. It is a handsome tree, cheering the beholder with its aspect of life and energy. When assembled in forests, they grow to the height of 80 or 90 feet, with a trunk 4 or 5 feet in diameter, entire two-thirds of its height.* In open situations, or in planted parks and rows, it stands 40 to 50 feet high, with a trunk one-third this height supporting a broad pyramidal leafy crown. But the aged trees assume a great variety of forms, picturesque or beautiful, which the artist is never weary of studying.

Analysis.—The *Roots* are often above ground, especially on the rocks they love, diverging many feet from the base, massive and strong, finally dissolving and descending deep. A cross-section of one will show the wood in annual layers inclosed in bark, but destitute of pith.

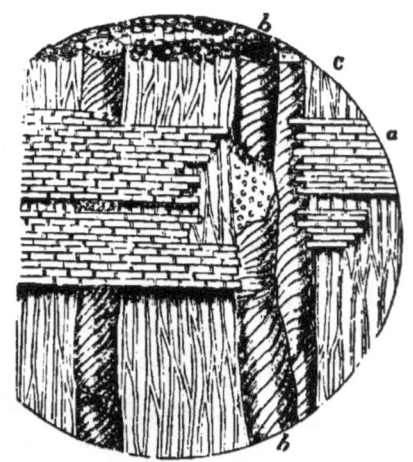

5, a shaving of the wood of Maple greatly magnified; *a*, the silver grain or medullary rays; *b*, spiral tubes conveying air or water; *c*, the proper woodcells.

The *Stem*, or trunk, in young trees is straight, erect, cylindrical, with bark slightly furrowed, gray, clouded with umber. With age it becomes shaggy with long, deep furrows in the bark, and angular with woody ridges from the main roots upward, and often bent and gnarly. The wood is hard, compact, pearly white, with a satin-like luster. Under a strong magnifier it appears as in the cut (5), showing clearly the three kinds of tissue of which it is composed.

---

* A tree in Blandford, Mass., 4 feet through at base, and 108 feet high, yielded seven and a half cords of wood,—*Emerson's Report*.

THE ROCK MAPLE.

The *Leaves* grow opposite, in pairs, on long, slender petioles, palmi-veined and reticulated. The blade is as broad as long, somewhat cordate at base, *extended with the veins* into 5 or 7 pointed lobes,* each bearing a few large teeth, and with rounded intervals between; smooth above, a little downy and pale-glaucous beneath. In the autumn, they undergo a wonderful change of color. From a bright green, of various shades in different trees, they become tinted and stained with the most brilliant hues—yellow, orange, scarlet, crimson, assuming often the very colors of flame, to the sudden alarm of the unwary.† (See illustrations, p. 295.)

The *Flowers* appear in April and May, together with the expanding leaves, proceeding from buds clustered at and near the end of the branchlets. They are yellowish green in color, in umbel-like corymbs, pendulous on slender, thread-like, downy pedicels about 2' long. There is a bell-shaped, fringed calyx with 8 or 10 stamens within, and no petals. In respect to fruit, the flowers are of two kinds. In the

---

* That infinite variety of beautiful and graceful forms for which the leaf is distinguished, becomes intelligible only when viewed in connection with its venation. Since it is through the veins alone that nutriment is conveyed for the development and extension of the tissue, it follows that there will be the greatest extension of outline in the direction of the largest veins. Pinni-veined leaves, wherein the midvein is the largest and all the rest side-branches, will generally be longer than wide, i. e., lanceolate, ovate, oval, oblong, oblanceolate, etc. Palmi-veined leaves, wherein there are several chief veins running from the base of the blade to the margin, will generally be broad in outline—as broadly ovate, or orbicular, or reniform; and often palmately trilobate, 5-lobed, 7-lobed, according to the number of veins. When the *veinlets* are comparatively weak, there may be a deficiency of tissue between the veins, causing the leaf to become either deeply lobed, or parted, or even divided up into several or many leaflets; in short, it thus becomes a compound leaf, either pinnately or palmately compound. Thus the student will notice with surprise that the general venation of a compound leaf differs in no wise from that of its corresponding simple leaf.

† The richest and most diverse hues that nature can produce by the separation and blending of all the prismatic colors, meet us in every grove, hill-side, and mountain. Red of every shade, from crimson to cherry; yellow, from bright sulphur to orange; brown, from clove-brown to liver-brown; and green, from grass-green to oil-green, stand forth in distinct spots, yet all mingled in fantastic proportions, clothing the landscape with an almost dazzling brilliancy, especially when lighted up by the mellow rays of an October sun.—*Hitchcock.*

sterile (♂), the stamens are prominently exserted and the stigmas deficient; in the fertile (♀), the stamens are deficient and hidden in the calyx, and the 2 stigmas prominent, with a double ovary.

The *Fruit*. As the ovary matures, a wing grows on the back of each carpel, converting the fruit into 2 winged *samaras*, or a double *samara* (a key), separable into two single ones. In each there is one seed, containing an embryo

6, section of a samara, showing the folded cotyledons at *e*; 7 to 11, progressive stages of germination.

with 2 large, folded cotyledons, and no albumen. It is instructive to watch the progress of these seeds in germination, as may be seen in all stages, in living specimens, under the Maples in Spring, as represented in the cuts.

The Sap of the Rock Maple is rich in sweetness, containing about 1 part of sugar to 30 parts of water. Early in March, or in February, while the buds are yet dormant, the sap begins to arise from the roots, and will overflow through tubes inserted in auger-holes cut deep into the wood for this purpose. The sugar is obtained by vaporization over hot fires. When the buds begin to open into leaves and flowers, the overflow of sap ceases.

The Name of the Rock Maple, *Acer saccharinum*, is characteristic—*acer*, sharp, vigorous, *saccharum*, sugar. Other kinds, both native and foreign, inhabit our forests and parks. (See *Botanist and Florist*, p. 74.) Among native species, *A. rubrum*, the Red, or Swamp Maple, with early crimson flowers and red-tinged leaves, will claim the learner's attention; also, *A. dasycarpum*, the White or Silver-leaved Maple, with leaves silvery-white beneath.

A. Pennsylvánicum, is a small, graceful tree, 12 to 20 feet high in northern forests, of many peculiar traits. Its leaves are generally 3-lobed, and the flowers with 5 petals, in long drooping racemes, are uncommonly showy. It is called Striped Maple, because of the smooth bark colored green and dark-brown in alternate longitudinal lines. It is the Moosewood in Maine, its bark and tender branches being the favorite winter food of the Moose; and it also bears the name of Whistlewood, from the facility with which the boys convert its straight, smooth branches into musical instruments.

LII. THE HORSE CHESTNUT.

Description.—This splendid tree is a native of Northern Asia, whence, by way of Constantinople, it emigrated to Europe, and from Europe to America. Here it is extensively planted for ornament and shade. It is noted for its rapid growth, massive foliage, and symmetrical proportions; but rejected by artists as wanting the picturesque.

Analysis.—The *Trunk* is a straight, cylindric column, with bark comparatively smooth, entire a third of its height, thence *excurrent,* giving off numerous straight branches and forming an oval or pyramidal, dense crown of foliage.

The *Leaves* are a perfect exemplification of symmetry and order. In arrangement they are in pairs, one leaf oppo-

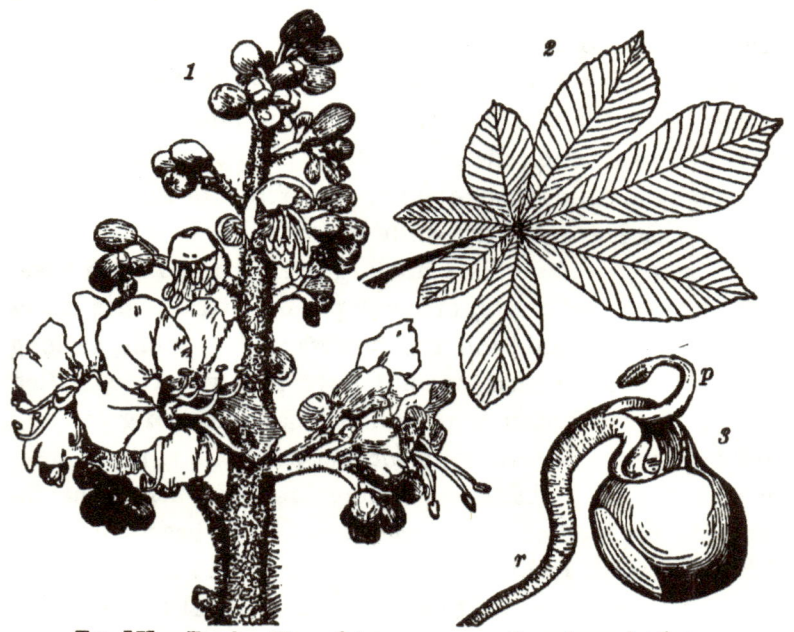

FIG. LII.—Æsculus Hippocástanum.   3, seed germinating.

site another,* supported on long slender petioles. At top the petiole divides, as in the Maple, into 7 veins diverging

---

* The learner has already observed three modes of leaf-arrangement, viz., the *alternate* in the Roseworts, etc., the *opposite* in the Maples, and the *verticillate* in the Loosestrife. He will now be interested in comparing them. In the alternate arrangement there is only one leaf at each node; in the opposite, there are two, and in the verticillate, 3 or more. The true nature of the alternate may be learned by an experiment. Select a straight, leafy shoot or stem of an Apple-tree, Evening Primrose, or any plant with seemingly scattered leaves. Beginning with the lowest leaf, fix a thread to the base of the petiole. Pass then, right or left, to the *next* leaf above and do the same; thence to the next in the same direction, and so on by all the leaves to the top. The thread will form a regular spiral. Let the same experiment be repeated in a shoot with opposite leaves, and two spirals running parallel with each other will be found; and in the case of verticillate leaves, as many spirals as there are leaves in each verticil. Hence the course of development in all growing plants is *spiral.* (See *Class-Book,* pp. 46–50, on *Phyllotaxy.*)

into a circlet, each becoming the midvein of a leaflet. The leaflets are inversely lanceolate, or *oblanceolate,* and serrate. Such leaf forms are *palmately compound* and *digitate* (finger-shaped), with the same venation as the simple leaf of the Maple (note, p. 193).

The *Inflorescence* is terminal, centrifugal, in showy, erect, pyramidal panicles, strongly contrasted in colors with the deep green of the foliage.

The *Flowers* are irregular, unsymmetrical, complete though often infertile. The 5 sepals united at base form a 5-lobed calyx. The 5 white petals dashed here and there with yellow and red, are entirely distinct. The 7 stamens with the 1 slender style are twice bent—downward, then upward. The ovary is 3-celled, with 2 ovules in each cell.

The *Fruit* is a 3-valved burr, beset with prickly points without, and occupied within by only one (rarely 2) large mahogany-colored seed.* It thus fails to fulfil the promise of its ovary. Of the 6 ovules, only one grows, to the suppression of the others and 2 of their cells. A careful examination will show the strangled rudiments.† The seed, often 1′ in diameter, includes 2 huge cotyledons inseparably united, without albumen. In germination, their 2 petioles (for the cotyledons are *leaves*) lengthen, and the plumule (the primary bud) issues from between them.

The Name, *Æsculus,* the title of this genus, was the ancient Latin name of a certain Oak with *esculent* fruit. *Æ. Hippocástanum* = horse-chestnut, alludes to its former reputation as a veterinary medicine.

Æ. glabra, with prickly fruit, and *Æ. flava,* with smooth

---

* One regrets that these beautiful seeds are not esculent like the Chestnut. They are however greedily eaten by deer, and in Switzerland they have proved to be an excellent food for sheep, giving a rich flavor to the meat.

† Similar *suppressions* habitually occur in the Oak, Birch, etc. The acorn is 1-seeded from a 3-celled, 6-ovuled ovary (p. 208-9).

fruit, are native species, called Buckeye. Both are large forest trees, with 5 leaflets and 4 petals. Other species are shrubs, with red or purple panicles, often seen in shrubberies.

Classification.—Æsculus and Acer would seem, at first view, to have little affinity with each other; but of late, botanists have included both, together with *Sapíndus* (Soapberry), *Staphylèa* (Stafftree), and other genera equally diverse in aspect, in the same order—the SAPINDACEÆ, or Soapworts. Their affinities are approximate rather than identical, so that the ordinal character cannot be satisfactorily formulated.

The **Soapworts** comprehend 73 genera, 650 species, divided into four suborders, found in all northern countries, and abundant within the Tropics.

Sapíndus (sapo-indicus = Indian Soap) gives name to the order. One of its species, *S. marginátus*, called Soapberry, grows in Georgia and westward. It is a small tree, with pinnate leaves, flowers in large panicles, and berries reddish-brown as large as grapes, and full of a soapy pulp. Other species in the W. Indies, more abundant in alkali, are actually used in washing linen.

Paullínia, of Brazil, affords the Guaraná, a popular beverage resembling tea in its effects. The seeds are dried, pulverized, kneaded into dough, then dried in cakes for the market.

LIII. SILK GRASS.

Description.—A stout herb a yard in height, surcharged with milk-white juice, and bearing globular clusters of bloom in June and July, is a sight familiar to the traveler in the low-lands along the streamlet or wayside. The plant is variously called Milkweed or Silk-grass. We shall leave the student alone, to study for record the organs constituting the leaf-region. The flowers and fruit present new and strangely curious structures.

SILK GRASS.

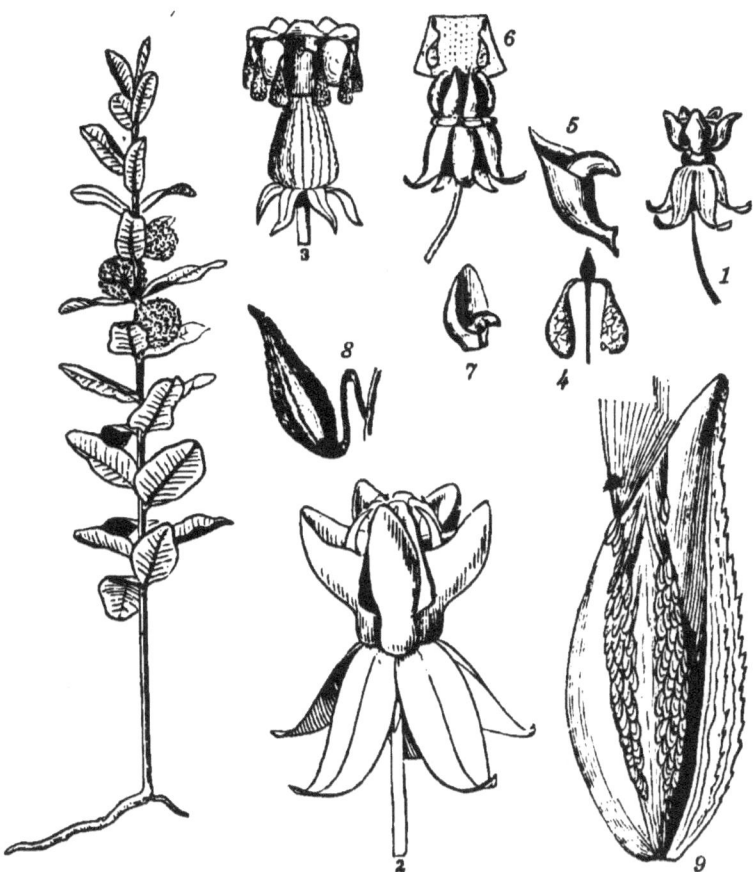

FIG. LIII.—Asclèpias Cornùti: 1, a flower natural size; 2, a flower enlarged; 3, the ovaries (advanced) and compound anther exposing the pollinia; 4, a pair of pollinia attached to the gland; 5, one of the hoods with its horn; 6, vertical section of anther and ovaries of A. phytolaccoides, with 2 pollinia in place; 7, a hood of the same; 8, a follicle; 9, a follicle open, showing the fledged seeds imbricated on the large placenta.

Analysis.—The *Inflorescence* is a simple umbel with pedicels (or *rays*) all of equal length and diverging in every direction. The 5 lanceolate petals, slightly *gamopetalous* at the base, are valvate in bud, and after opening, sharply reflexed, concealing the 5 sepals, and exposing the *corona* (staminate crown) to view. This consists of 5 fleshy, rose-white *hoods* attached to the mass of united anthers and stigmas. From the opening of each hood projects a little curved horn. Both hood and horn are of unknown use.

Pollen. Of the 5 anthers, each contains 2 club-shaped masses of pollen (*pollinia*) suspended in pairs beneath the disk of the stigma by slender stipes attached to 5 double glands. The pollinia of adjacent anthers are so united. The glands are very sticky and adhere to such insects as call in quest of honey, while their pollinia are dragged out of the anthers and carried to other flowers. This may be for the purpose of cross-fertilization; but the double pollinia dangling "like saddle-bags" from the legs of the insects often prove very annoying.

Under the staminal mass are 2 ovaries, each 1-celled with numerous ovules. But few of the ovaries in the umbel are fertilized and come to maturity.

The *Fruit* is lance-shaped, with a rough exterior, 1-celled, and opens by a slit along the inner side. Such we call a *follicle.* It incloses many flat seeds imbricated on the large placenta, each fledged with a tuft of long silky hairs called a *coma.* These serve, like the down of the Dandelion, to waft the seeds to a distance.

The Scientific Name of the Milkweed shown in the figure, is *Asclèpias Cornùti,* or the Horned Asclepias, the genus being dedicated to Æsculapius, the god of Medicine. There are 50 species, which differ, however, in only a few particulars from the description in the text.

Classification.—The order of the Asclepiads (ASCLEPIADACEÆ) may be formulated as follows :

    Plants with a milky juice.
    Flowers regular, perfect, 5-parted, symmetrical.
    Stamens and stigma consolidated.

SILK GRASS.

Anthers, each with 2 pollinia.
Ovaries 2, with 1 stigma.
Fruit, 1 or 2 follicles.
Seeds with a coma.

ORGAN.	*L*ife, *H*abit, *N*umber, *P*lace, *D*ehiscence, *K*ind, *C*onstruction, *F*orm, *P*lacentation, *S*ize, *Q*ualities, *A*ppendages.
Plant, L.H.S.Q.	♃, *erect, branching, 2–3 ft., leafy, pubescent in lines, milky.*
Root, L.Q.	♃, *axial, branching.*
Stem, L.H.K.F.	*Cauls erect, herbaceous, branched, 4-sided.*
Leaves, L.P.C.F.S.Q.	*Petiolate, opp., pinni-v., lanceolate, pointed, obtuse at base.*
Inflorescence, P.K.A.	*Umbels term. and axillary, pedunculate.*
Flower, N.C.	∞, *perfect, 5-parted, gamop., with a corona of 5 hoods*
Calyx, F.Q.	*Small, rotate, valvate.* [*seated on the stamens.*
Sepals, L.N.P.F.	*Decid. 5, spreading, ovate, smooth.*
Corolla, F.Q.	*Gamopetalous, valvate, rose-red.*
Petals, L.N.P.F.	*Decid. 5, oblong, reflexed hoods shorter than the slender*
Stamens, N.P.C.	*5, united, on the corolla at base.* [*incurved horns.*
Anther, D.C.F.	*Each vertically 2-celled, joined to the stigma.*
Style, N.C.F.	*None.*
Stigma, N.F.	*Pollinia united in pairs to 5 sticky glands on the 5 angles*
Ovary, C.F.Pn.	*2, distinct, conical.* [*of the stigma.*
Fruit, N.D.K.F.Q.	*Follicle mostly but one, smooth.*
Seed, N.C.F.Q.A.	*Anatropous, flat, imbricated, oval, comous.*

LOCALITY.—*In a swamp, Lexington, Mass.* (Date), *June 20, 1870.*
CLASSIFICATION.—GAMOPETALOUS EXOGENS
   —Order, ASCLEPIADACEÆ, THE ASCLEPIADS.
NAME.—Latin, **Asclepias incarnata.**
 —English, *Swamp Milkweed.*
REMARKS.—*The corona is rose-red like the petals.*

The **Asclepiads** number 146 genera and 1300 species, most abundant in S. India, S. Africa, and Australia. There are comparatively few species in the United States.

Butterfly-weed, or Pleurisy Root (*Asclèpias tuberòsa*), a handsome plant with orange-colored flowers, native in our pastures and meadows, is employed medicinally as a laxative and diaphoretic.

Dischídia, of E. India, is a famous Pitcher Plant.

The Cow Tree of Ceylon (*Gymnèma lactiferum*) yields a bland, wholesome milk which the natives use for food.

The Wax Plant (*Hoya*), from the W. Indies, with wax-like leaves and umbels, is a favorite house plant.

Stapèlia, with flowers so fœtid as to deserve the name "Carrion Flower," is a large S. African genus.*

The Record of *A. incarnàta*, another species quite common, is here annexed as a model for the order.

Scientific Terms.—Coma. Corona. Gamopetalous. Hoods. Horns. Pollinia.

LIV. SPOTTED KNOTWEED.

Description.—In June, and after, the Spotted Knotweed displays its flesh-colored spikes. Like the other foreigners, it seeks cultured fields and the waste corners about our dwellings; and the garden which is free from its encroachment is well kept. None favors the intruder, yet the botanist may profit by the study of its wonderful organization.

Analysis.—The *Root*. We first note that the root is *axial* in its kind, and a cross-section shows but one woody layer; hence it is annual. The stem is remarkable for the distinctness of the internodes, the nodes being excessively swelled, and *booted* with the stipules.

---

* This plant is sometimes cultivated in the green-house for the sake of its grotesque branches and pretty flowers. So carrion-like is its odor that the common blue-bottle fly is said often to make the mistake of "blowing" it, i. e., of depositing its eggs upon the petals, where they occasionally hatch, but only to starve.

SPOTTED KNOTWEED.

Fig. LIV.—Polygonum Persicària : 2, portion of a cluster enlarged ; 3, a flower; 4, ovary and 2 styles ; 5, achenium ; 6, seed dissected, showing the embryo.

The *Stipules* are of a pattern called *ochreæ*. They grow in pairs from the base of the petiole as usual, but unite into a membranous sheath clasping the node and stem like a boot (ochrea), and in this species are fringed, or *ciliate* with a few long hairs. The outline, margin, construction and quality of the leaf, including the heart-shaped spot in the center, should all be noted.

The *Flowers*, small and numerous, are supported on pedicels; hence the cluster, which seems from its density a spike, is properly a raceme. They are regular, but very unsymmetrical, consisting of 5 sepals, 6 stamens, 2 stigmas and 1 ovary. Like the flowers of Hepatica (p. 55) they are *apetalous*, having but one set of envelopes.

The *Fruit*. The ovary ripens into a lens-shaped, black, polished achenium still inclosed in the persistent, rose-colored calyx. The one seed contains a curved, inverted embryo on the side of a starchy albumen.

The Name.—*Polýgonum Persicària* is the classic name, the former meaning many-jointed (Gr. *polys*, many, *gonè*, joints); the latter, peach-leaved, alluding to the resemblance of the leaves to those of the Peach-tree (*Pérsica*, Tournef.). Other species of Polygonum will also be found flowering in June and July, and may be profitably studied with this, to mark the distinctive specific characters of each, viz.:

P. Pennsylvánicum, the Pennsylvanian Knotweed, a native species growing in wet places, has the upper parts beset with minute glandular hairs (glandular-hispid), the flowers in dense racemes, stamens 8, etc.

P. aviculàre, the Bird Knotweed, prostrate in dooryards, has small (1' and less) leaves, and minute axillary flowers. The seeds furnish food for many wild birds.

P. amphíbium (amphibious) grows either in water, or on land. It is our largest native species, with leaves 5–7' long and bright red flowers in thick spikes 1–2' long. Stamens only 5. It is smooth when growing in water, viscid-hairy, on land.*

---

* The beautiful rosy petals of *Polygonum amphibium* are rich in honey. The stamens, however, are short, and the pistil projects above the corolla. The nectar is unprotected and accessible even to small insects like the ant. The stamens ripen before the pistil, and any flying insect, however small, coming from above would assist in cross-fertilization. Creeping insects, on the contrary, would rob the honey

SPOTTED KNOTWEED.

ORGAN.	Life, Habit, Number, Place, Kind, Construction, Form, Placentation, Size, Qualities, Appendages.
Plant, L.H.S.Q.	☉, *herb, in damp places, 1-2 ft., glandular-hispid above.*
Root, L.K.	*Annual, axial, branching.*
Stem, L.H.K.F.	*Herbaceous, erect, branching, with nodes swollen.*
Leaves, L.P.C.F.S.Q.	*Alter., pin.-veined, ochreate, lanceolate, rough-edged, 2—5'.*
Inflorescence, P.K.A.	*Terminal, racemes, pedunculate.*
Flower, N.C.	*Unsymmetrical, perfect, apetalous, 1" diameter.*
Calyx, F.Q.	*Polyphyllous, rosaceous, rose-colored.*
Sepals, L.N.P.F.	*Persistent, 5, imbricated, erect, oval.*
Corolla, F.Q.	*Wanting.*
Petals, L.N.P.F.	*Wanting.*
Stamens, N.P.C.	*8, hypogynous, filament slender, included.*
Anther, D.C.F.	*Innate, longitudinal, 2-celled, oval.*
Style, N.C.F.	*Two-parted, terminal.*
Stigma, N.F.	*Two, terminal, capitate.*
Ovary, C.F.Pn.	*Double, superior, ovoid.*
Fruit, N.D.K.F.Q.	*1, indehiscent, achenium, lenticular with flat sides.*
Seed, N.C.F.Q.A.	*1, albuminous, dicotyledonous.*

LOCALITY.—*Ditches, Terre Haute, Ind.* (Date), *June 12.*

CLASSIFICATION.—APETALOUS EXOGENS.

—Order, POLYGONACEÆ.

NAME.—Latin, **Polygonum Pennsylvanicum L.**

—English, *Pennsylvanian Knotweed.*

REMARKS.—*The upper nodes and peduncles rough or hispid with minute stalked glands.*

without benefiting the plant. To prevent the visits of the latter, therefore, the hairs secrete a viscid fluid, which makes the stem slippery and difficult to climb. This plant, as its name denotes, may live in the water. In that case it is safe against those climbing pilferers, and then the stem is smooth, with no hairs and no sticky substance. The arrangement is a special one, and furnished only when needed in the economy of the plant.

P. orientále, Lady's Thumb, a foreigner, about houses, tall (6 ft.) and stout, has the ochreæ with a spreading border, the flower clusters large, rose-colored, stamens 7, etc.

Classification.—The order POLYGONACEÆ—the Sorrelworts—may be characterized as follows:

Herbs with alternate leaves and swelled joints.
Stipules in the form of ochreæ sheathing the stem (a feature by which the order may be recognized at sight).
Flowers apetalous, with a persistent calyx.
Ovary 1-celled, with 2 or 3 styles or stigmas.
Achenium with 1 erect albuminous seed.

The Sorrelworts number 33 genera and 690 species abounding in all countries. Among these are—

The Buckwheat Plant (*Fagopyrum*) indigenous in Northern Asia, now extensively cultivated as an article of food in general use, and by bee-keepers as a valuable honey plant. The small black kernel with white albumen, whence the 'flour' is obtained, has, as every one knows, the form of a Beechnut (German, *Buch*). Hence its name, both English and classical, is equivalent to Beech-wheat.

Rhubarb (*Rheum Rapónticum*), also from Asia, is the well-known Pie Plant. The pulpy tissue of the petioles is made acid by the oxalate of lime. Several species of Rheum yield the medicinal rhubarb-root of the shops.

Dock (*Rumex*), both the Broad-leaved and the Narrow-leaved, everywhere abounds as a "pernicious weed;" yet the roots of some species afford a valuable medicine.

Sheep Sorrel (*Rumex Acetosélla*) has a pleasant acid foliage, and abounds in old fields and pastures where there is a lack of alkali in the soil, reddening with its minute flowers many a sterile knoll and hillside.*

Scientific Terms.—Apetalous. Ciliate. Ochreæ. Raceme. Spike.

---

* In their modes of fertilization there is much diversity among the plants of this order. The various species of Rumex are destitute of honey, and wind-fertilized. Of the Polyganums, *P. aviculáre*, the Bird Knotweed, is probably self-fertilized. *P. Persicaria* is proterandrous, its stigmas ripening after their anthers have shed their pollen, while the Buckwheat is *dimorphous*, some of it with long stamens and short styles, others with long styles and short stamens.

LV. THE SPURGES.

Description—Some of these homely plants are common throughout the country. They are noted for their acrid, milky juice. The attention of the botanist is due them on account of the strange construction of the flowers. The Spurge here figured will be found blossoming in June and after, in open fields and waysides.

Analysis.—GENERIC CHARACTERS.—The Flowers of the Spurges are often too small to be understood without the aid of a microscope. The "calyx" is cup-shaped, bearing on its margin 5 or 4 glands of peculiar form and red or white color. Within it stand several or many stamens, each with a minute bractlet attached at its base, and a joint above. In the midst, is an ovary raised on a foot-stalk and tipped with 3 styles, each 2-cleft, so that there are 6 stigmas (half-stigmas).

Now what mean these bractlets, joints, and foot-stalk? They imply, as botanists interpret, that each stamen is a flower of itself—a staminate, monandrous flower with a pedicel in the axil of a bract; that the ovary is a pistillate flower consisting of 3 united carpels; and the "calyx" is an involucre inclosing the little flower-group. As it grows older, the pistillate flower arises on its pedicel quite outside of the involucre and ripens into 3 carpels, separable into 3 nutlets, each with one seed.

The milk-white juice already noted, flows from every incision, is always acrid in taste, in some species venomous, and it should be avoided.

The Name *Euphorbia* is the title of the genus characterized in the above description—a genus of vast extent, growing in all countries and embracing more than 700 species. The original one (*E. officinàrum?*), discovered by King Juba in Barbara, was so named by him in honor of Euphorbus, his chief physician.

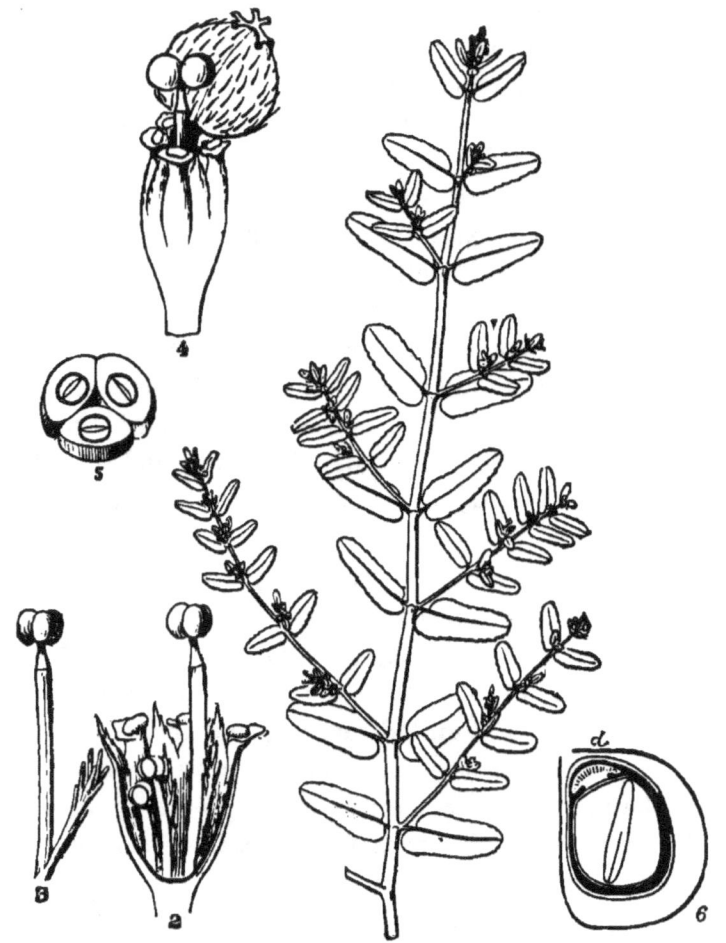

Fig. LV.—Euphórbia maculàta : 2, section of an involucre showing the ♂ flowers; 3, a ♂ flower with its bract ; 4, an involucre entire, showing the 1 ♀ flower, etc.; 5, section of ovary ; 6, section of a seed of *E. Láthyris*, with embryo, and (*d*) caruncle.

SPECIFIC CHARACTER.—The species before us differs from all others in the following combination of characters : Root annual. Stems prostrate, diffuse, reddish, puberulent, with opposite leaves and alternate branches. The leaves are of two sizes, 3″ to 6″ long, oblong, very oblique, obtuse, serrulate, with a red-brown spot in the center, and small fringed stipules at the base of the short petiole. The minute flowers issue in dense, bracted, lateral clusters on a short peduncle, making no display. Glands of the involucre 4, red. Seeds ovoid, 4-angled, transversely *rugous* (wrinkled), with no caruncle, as some species have (6, *d*). This is *E. maculàta*, the Spotted Spurge.

E. hypericifòlia is another closely related and equally common species. It differs only in being erect (1-2 ft.), with leaves larger (1′), often slightly *falcate* (curved like a sickle $=falx$), and the flowers terminal.

E. corollàta, abundant westward, is our most showy kind. It stands erect 2-3 feet, bearing an umbel of white 5-lobed involucres. Its perennial root is a purgative more violent than Ipecac.

Classification.—The order EUPHORBIACEÆ (the Spurgeworts) is very large, generally limited as follows:

> Plants with a milky, acrid juice.
> Flowers incomplete and imperfect.
> Ovary free, 3-celled, with 3 or 6 stigmas.
> Ovules suspended from the top of the cell.
> Fruit 3-lobed, separating into 3 carpels.
> Seeds 1 or 2 in each carpel, anatropous.
> Embryo straight, 2-lobed, in oily albumen.

The Spurgeworts number 190 genera, 3200 species. As a whole, the milky juice is venomous, but many species afford valuable oils, resins, and farinaceous food.

Castor Oil is expressed from the seeds of *Ricinus communis*, a well-known gigantic annual in Northern gardens, but a stately tree in the South.

Croton Oil, a powerful purgative and external irritant, is from the seeds of *Croton Tiglium* of India.

The tonic Cascarilla is the bark of *Croton Eleutèria* of Brazil.

Capers, used in pickles, sauces, etc., are the 3-lobed fruit of *Euphórbia Láthyris*, often seen in our gardens.

Tapioca is obtained from the Bitter Cassava (*Jatròpha Mánihot*), a shrub extensively cultivated in S. America. Its tuberous root, sometimes weighing 30 lbs., is full of a poisonous juice. In preparing it for food, it is first scraped to a pulp and pressed to remove the poison. The cakes of cassiva thus formed are dried and baked, making a bread commonly used by the poorer classes. When the expressed juice is allowed to stand, a delicate starch is deposited, which, when washed and granulated on hot iron plates, forms the Tapioca of commerce.

India Rubber is the thickened juice of *Siphònia elástica*, a tree growing in Guiana (see *Chemistry*, p. 227).

Boxwood, used by engravers, and for mathematical instruments, and also cultivated for borders, is *Buxus sempérvirens* of Asia Minor.

The so-called Blinding Tree (*Excœcària Agállocha*) of the Moluccas has a juice so acrid that a drop falling into the eye will nearly blind it—an accident which is said to have happened to sailors sent on shore to cut fuel. Even the smoke of the burning wood is dangerous.

LVI. THE WHITE OAK.*

"Not a prince
In all that proud old world beyond the deep
E'er wore his crown as loftily as he
Wears the green coronal of leaves with which
Thy hand hath graced him."
BRYANT.

Description.—A large proportion of our forest trees are Oaks. Also in the open fields the Oaks stand solitary, in alternation with Elms and Maples, the charm of every rural scene. The White Oak will be our special theme to-day. Its flowers appear in May, soon after the expanding leaves. The

---

* The Oak, Pine, etc., are fertilized by the wind. It is curious to notice, in contrast with the insect-fertilized plants we have considered, the new floral adaptations which here exist. The long, lightly-hung, pendulous catkins are set in motion by the merest breath of air. The blossoms appear, too, in the early season when gales are most numerous and boisterous,

THE WHITE OAK.

blossoms and fruit of any Oak will, however, serve for this lesson.

Analysis (generic).—The Oaks put forth two kinds of flowers on the same tree. The sterile or staminate (♂) are

FIG. LVI.—Quercus alba.

disposed in long, slender, pendulous clusters called *aments* or *catkins*, several from one bud. They consist merely of a 5-8-lobed calyx with 5-8 stamens. The fertile or pistillate (♀) are solitary, or few together—each an ovary with 3 stigmas invested with a scaly involucre. The ovary is 3-celled, with 2 pendulous, anatropous ovules in each cell.

QUERCUS. 209

But in ripening, only 1 of the 6 ovules becomes a seed. By its fruit—the *acorn*, the Oaks are universally known. It is, by *suppression*, a 1-seeded nut partly immersed in a scaly, cup-form involucre. On dissection, we find in the seed an embryo with 2 massive cotyledons, the short radicle pointing upward, destitute of albumen.

Germination. Under the Oaks at the time of flowering, the student will find acorns of the preceding year in all stages of germination as shown in the cuts. The swelled cotyledons (which are but transformed leaves) cannot extricate themselves from the shell, but burst it and thrust forth their petioles with the radicle and plumule between them, the former to grow downward, the latter upward.

The Name of this noble genus is the classic Latin one — *Quercus.** In the United States there grow as many as 25 species, and at least 6 or 8 in every vicinity. The practiced eye will distinguish them by their tree-forms alone. All may know them by the forms of their leaves (Appendix). To identify them by verbal description is

Acorn (seed of *Quercus palustris*) germinating: 6, section showing the radicle (*r*) which is to become the root, and the two cotyledons (*c*) which are to nourish it; 7, the radicle *r*, descending; 8 and 9, the radicle *r*, descending, and the plumule (*p*) ascending.

---

* The Oak has been identified with man's history from the earliest ages. Its groves have been held sacred alike by Jews (Gen. xxi, 23), Greeks, Romans and Celts. The

often difficult, and a test of scholarship. (See *Bot. and Flor.*, p. 305.)

SPECIFIC CHARACTERS of the White Oak (*Q. alba L.*). This tree is known at sight among its compeers by its light ash-colored bark breaking into square loose flakes on the surface. The leaves on short petioles are deeply divided into obtuse segments, 3 or 4 on each side, none angular, all bounded by flowing outlines. From a bright green they change to violet and purple in Autumn, and many are persistent. A new feature in venation is here to be noticed. The leaves are *straight-veined*—the veinlets continue straight through the blade to the margin. Compare the leaf of Beech, Chestnut; also of the Apple. The stipules are *fugacious*.

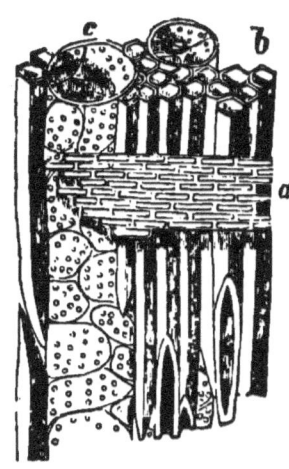

Wood of Oak, greatly magnified: *a*, medullary rays; *b*, wood cells; *c*, ducts.

The *Acorn* ripens in the Autumn following its flower; is nearly sessile, 1' long, an ovoid nut one-third immersed in a hemispherical cup. The seed is well-flavored, and eaten by man as well as beast.*

Oak was consecrated to Jupiter, even to a proverb; and the Druids (*drus*, an Oak) are supposed to have been named from their superstitious regard for the Oak and the Mistletoe which grew upon it. The Greeks adopted it as the emblem of hospitality. In Rome, to obtain a "crown of Oak," it was necessary to be a citizen, to slay an enemy, to save the life of a Roman, or to reconquer a field of battle.—The Oak is, however, peculiarly a British tree, associated with English naval victories—with the "Walls of Old England" and the "hearts of Oak" that have beaten bravely within them. Many an Oak has become historic; like the Oak of Torwood, within whose hollow slept the famous Wm. Wallace; the Royal Oak that sheltered the fugitive Charles after the battle of Worcester; Pope's Oak in Windsor Forest; while in this country we recall the Charter Oak of Hartford.

* Some species of Oak, as Red Oak (*Q. rubra*), Pin Oak (*Q. palustris*) are *biennial-fruited*; i. e., they require 2 years from flowering for their acorns to ripen.

In England, whose Oak forests are now valued for timber, some centuries ago the Saxons valued them only for their acorns, or *mast*, on which their swine were fat-

QUERCUS. 211

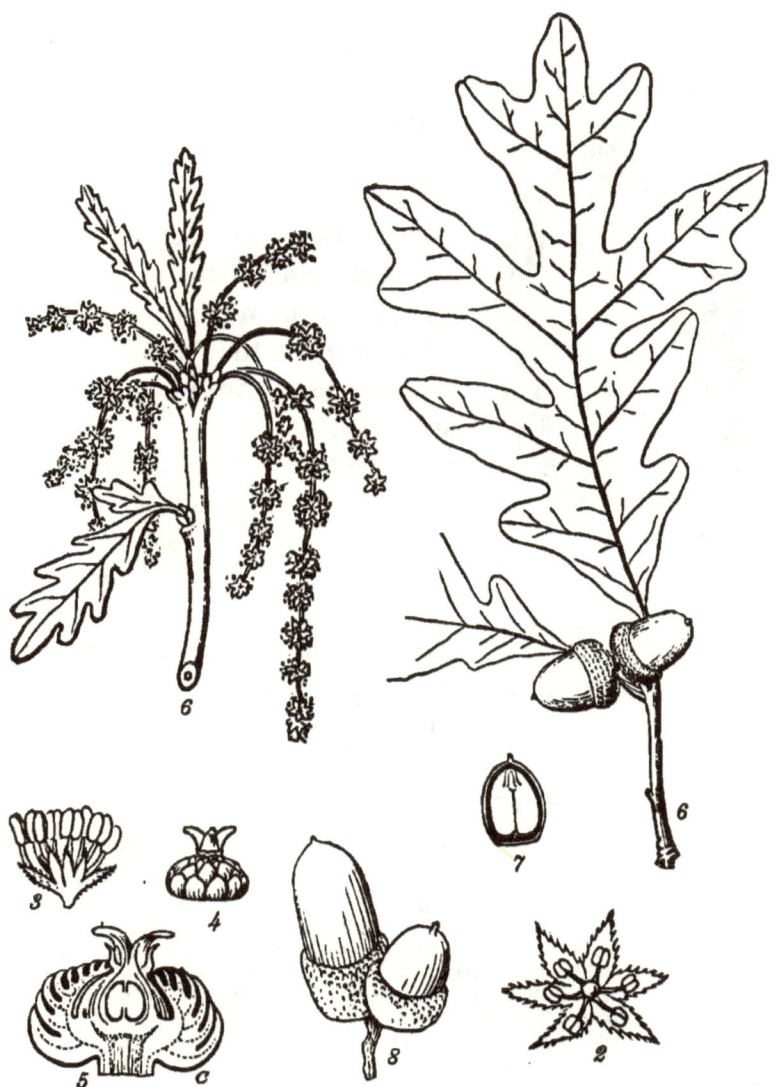

6, young branchlet of Q. alba, with aments, &c. 2, a staminate (♂) flower; 3, the same; 4, a pistillate (♀) flower with 5 stigmas; 5, vertical section of the same; 6, branchlet with full-grown leaves and mature fruit; 7, section of the the acorn showing the two thick cotyledons and embryo at top; 8, acorns of Q. robur.

THE WHITE OAK.

ORGAN.	Life, Habit, Number, Place, Kind, Construction, Form, Placentation, Size, Qualities, Appendages.
Plant, L.H.S.Q.	*Tree deciduous, in forests, 2 ft. diam., 60 ft. high.*
Root, L.K.	*Axial, branching, extensive.*
Stem, L.H.K.F.	*Arboreous, exogenous, erect, trunk terete.*
Leaves, L.P.C.F.S.Q.	*Decid., cond., alt., straight-veined, lanc., serr., acum.*
Inflorescence, P.K.A.	*Axillary, ♂ long catkins, ♀ 3 together in a burr.*
Flower, N.C.	*6-parted, apetalous, monœcious, small.*
Calyx, F.Q.	*Cup-form, 6-parted, cream-greenish.*
Sepals, L.N.P.F.	*Persist., 6, ♀ adherent, erect.*
Corolla, F.Q.	*None.*
Petals, L.N.P.F.	*None.*
Stamens, N.P.C.	*8-20, filiform, showy, erect.*
Anther, D.C.F.	*Oval, longitudinal, 2-celled, versatile.*
Style, N.C.F.	*3, united at base, short.*
Stigma, N.F.	*3, club-shaped (9 in the involucre).*
Ovary, C.F.Pn.	*3-celled, 6-ovuled, ovoid, or plano-convex.*
Fruit, N.D.K.F.Q.	*Mostly 3 nuts in a burr (prickly involucre).*
Seed, N.C.F.Q.A.	*1, anatropous, white, cotyledous, farinaceous.*

LOCALITY.—*Chesterfield, N. H., woods.* (Date), *May 19, 1836.*

CLASSIFICATION.—APETALOUS EXOGENS.

    ORDER.—CUPULIFERÆ, or MASTWORTS.

        NAME.—Latin, **Castanea vesca Linn.**

          —English, *Chestnut.*

REMARKS.—*Fruit sweet and nutritious, falling in October. Timber light, very durable.*

tened. The right of feeding hogs in the woods, called *pannage,* became a valuable kind of property. With this right Monasteries were endowed, and it often formed part of the dowry of the king's daughters. To regulate and secure these rights, rigid laws were enacted and records kept. When William the Conqueror converted the New Forest into a hunting-ground, the anger of the people was due to the loss of food for their droves of swine.

In this connection, let the student analyze the Chestnut Tree, the Beech, or the Hazel. A sample tablet is annexed.

Classification.—The order CUPULIFERÆ—the Mastworts—is thus limited to

> Trees or shrubs with very deciduous stipules.
> Leaves alternate, simple, straight-veined.
> Flowers apetalous, monœcious, the ♂ in catkins.
> Ovary adherent, with all but *one* cell and ovule abortive.
> Fruit a nut, one or more together in a cup or sack.
> Seed one, filled by the embryo with its massive cotyledons.
> Albumen none.

The Mastworts number 8 genera and 250 species. Among them are the Oaks, Beech,* Chestnut, Iron-wood, Hazel, etc., important for their timber and fruit.

Chestnuts are the fruit of *Castánea vesca*.† The American variety is smaller and sweeter than the Spanish Chestnut of Europe. Beech-nuts, the fruit of our *Fagus ferrugínea*, are very sweet and nutritious. Filberts, the fruit of the Hazel (*Córylus*), come from Europe. Our own Hazel-nut is nearly as good. The acorns of the White Oak and Chestnut Oak (*Q. Prinos*) are eatable.

Nutgalls are produced on the leaves and twigs of Oaks by the puncture of insects depositing their eggs. The nutgalls of commerce used in making ink, etc., come from Asia Minor. They abound in tannic acid, a principle also found in the bark of some species of Oak used in tanning leather.

The timber especially of the Live Oak (*Q. virens*), White Oak, and English Oak (*Q. robur*), is of great value in shipbuilding and all

---

* To the German name of the Beech (buch) we owe our English word book, the sides of thick books having formerly been made of beech boards.

No tree of the forest has its tint of trunk more varied by mosses, lichens and handsome kinds of fungus that always diversify its dark-gray bark. Virgil loved a Beech-tree for the abundant shadow it gave him, and Gray wandered to be soothed among the famous Burnham Beeches, which he says "are always dreaming out their old stories to the winds."

† In parts of Europe the Chestnut is highly valued as an article of food, and the tree is extensively grown for this product alone. Many centuries ago Martial said:
> "For Chestnuts roasted by a gentle heat
> No city can the learned Naples beat."

The Chestnut is yet roasted daily there as well as in many other Italian cities; and similar scenes are enacted on our own street-corners. In the south of France it forms the common vegetable diet of the peasantry.

mechanic arts where toughness, strength, and durability are requisite. The wood of Chestnut is eminently durable; that of Beech, Ironwood (*Carpínus*) and Liver-wood (*Óstrya*), is hard and compact, and therefore serviceable for joiners' tools.

LVII. THE WHITE PINE.

Description.—The White Pine grows in any soil where it is planted; but its native forests and groves are generally associated with a dry sandy loam. Our Pilgrim Fathers found here one continuous forest waving with Pines, where now are cities, towns and plantations. On the plains of Dartmouth and Saratoga once towered majestic Pines more than 200 feet. To-day, on the Sierra Mountains, the Lambert Pines 300 feet in height lift their imperial heads.

Analysis (generic). The *Leaves* of the Pines are truly evergreen, persisting in all their verdure through the Winter until those of the next season are full grown. Their form is as characteristic as that of the cones. They are *acerous* or needle-shaped, angular, collected in little *fascicles* (bundles) of 2s, 3s, or 5s, bound together by a sheathing bract at the base. In 2s they are semi-terete; in 3s and 5s triangular, with serrulate edges.

The *Flowers* come with the new leaves. They are of two kinds, both generally found on the same tree, i. e., *monœcious*. The sterile (♂) flowers are in small, oblong, dense, reddish aments clustered around the base of the new shoots. Each ament is involucrate with a few scales, and consists of stamens alone. The anthers are 2-celled and contain *triple* pollen grains. The fertile (♀) aments are lateral, consisting of spirally imbricated scales (open carpels) each bearing at its base 2 ovules turned downward, although not inverted on their stalks (orthótropous).

The *Fruit* is not matured until the second year after its flowers (*biennial*). It is then a *cone* formed of the grown and hardened fertile ament, with its scales generally thick-

Fig. LVII.—Pinus Strobus, a young tree and grove.

ened at the edge, at last relaxed and spreading, freeing the 2-winged seeds nurtured in the lap of each. The student will not fail to notice the total absence of a style or stigma;

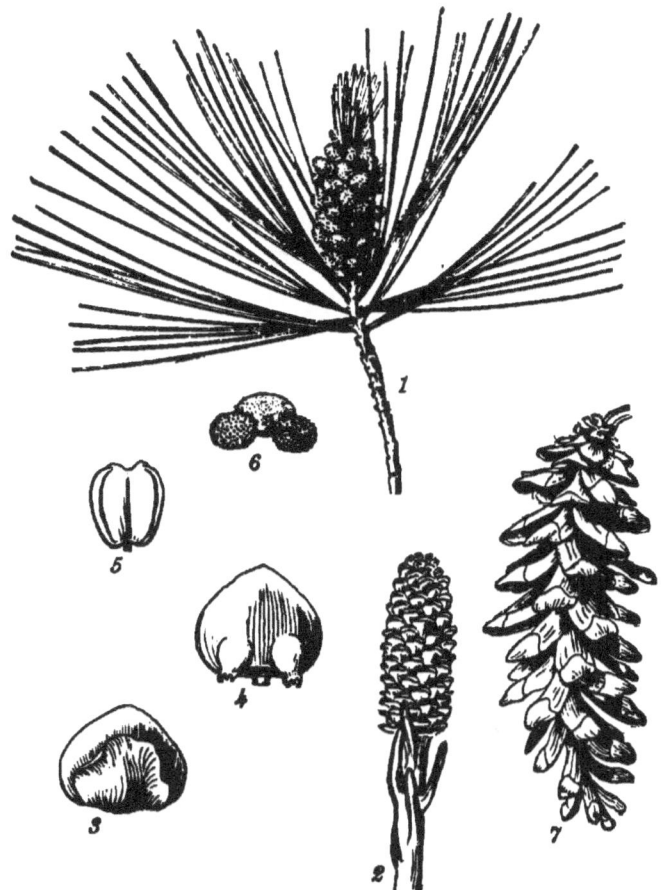

Pinus Strobus : 1, a branchlet with staminate flowers ; 2, branchlet with pistillate flowers ; 3, a carpellary scale with its bract ; 4, the same seen from within, with its 2 ovules turned downward ; 5, an anther ; 6, a grain of pollen (triple) ; 7, a ripe cone with its scales relaxed.

neither is there any proper ovary or seed-vessel. The carpellary scales which should invest the ovules and seeds, only subtend them ; hence they are truly *naked*.* The embryo, resting in oily albumen, has 3–12 cotyledons.

---

* Fertilization is effected by the direct application of the pollen to the ovule instead of to an intervening stigma. The wind is the agent for conveying the pollen to its place. (See Note, p. 31.) There is therefore no need of attracting insects by brilliant colors and pleasing perfumes; hence the flowers are inconspicuous and inodorous.

The Name of this grand and useful genus is the ancient Latin term—*Pinus,* from the Celtic *pin* or *pen,* a rock or mountain. The White Pine, that species to which our figures chiefly refer, is *Pinus Stròbus*—the "Weymouth Pine" of the English parks. This is the tallest of all our forest trees, many with a diameter of 4 or 5 feet, rising to 100 and 140 feet. The trunks perfectly straight, erect, free from limbs, extend ¾ their whole height, affording a strong, soft, light, and durable timber, more extensively used in architecture than any other kind.

SPECIFIC CHARACTER.—The *Root* of *P. Strobus* penetrates the soil but 2 or 3 feet, and is quickly dissolved into irregular branches and branchlets, filling a space of 30 to 40 feet diameter.*

The *Trunk* is cylindric, erect, with a smooth bark in trees less than a foot in diameter and in old forest trees regularly broken into long narrow plates. The branches are given off in whorls and at nearly right angles, one new whorl each year. In forests, all but the upper branches soon perish, and these stretching out over the other trees render the Pines conspicuous in the distant landscape.

The *Leaves* are in fascicles of 5s, and 4' in length.

The *Cones,* nearly 6' long when ripe, have scales slightly if at all thickened at their edges, thus quite unlike the other Pines. Compare this with—

P. rígida, the Pitch Pine, which has its leaves in 3s, cones ovoid, with scales thick-edged and *clawed* at the end, and bark rough and black, a tree 30 or more feet high.

P. resinòsa, Red Pine, has leaves in 2s, cones ovoid-conical,

---

* The roots of the White Pine are almost incorruptible. In clearing up new lands where the Pines have been felled or blown down, the stumps with their roots are often taken up and used in making a fence, by setting the under surface of the roots to form the outer or the finished side. Fences so made exhibit, after a hundred years, few signs of decay.—*Emerson.*

the scales not claw-tipped, bark rather smooth, tree 50–80 feet high; both species native northward.

P. palústris, Long-leaved Pine, in the lowland forests of the South, has leaves in 3s, and 10–15′ long and cones of nearly equal length.

LVIII. THE HEMLOCK.

Description.—The Hemlock grows in the forests of all the States west to Oregon, especially loving a granitic soil; and in Canada and New England, the tree and its products are so common that Hemlock is almost a household word. Flowers and fruit (last year's cones) may be found in May.*

Analysis.—GENERIC CHARACTERS.—The leaves are solitary (not fascicled), short, of one kind, and persistent two years. The trunks are of that class called *excurrent*—running distinct through to the summit of the pyramidal head. Here also we have flowers of two kinds (monœcious), both in aments, and on the same tree. Mark their situation, not on the new shoots, as in the Pines, but on the branchlets of the preceding year; the ♂ aments in the axils of the upper leaves; the ♀ terminal. The cones mature in the Autumn of the first year. Their scales are thin-edged, never embossed nor clawed, each 2-seeded, and subtended by a bract.

The Name given to the genus possessing these traits is *Abies*—the ancient Latin for Spruce. It comprises the Spruces, Firs, and Hemlocks, evergreen, resinous trees, like the Pines except in the above obvious distinctions.

SPECIFIC CHARACTERS.—The Hemlock when young has a peculiar grace both of form and foliage. With age, it becomes rugged and unsightly. In forests, the trunk is

---

* Specimens of our native Spruce or Fir, or of the Norway Spruce, so common in our parks and door-yards, will answer for this study.

often sixty feet high, beset above with knots among its scragged branches. The leaves are short-linear, silvery beneath, on delicate petioles, spirally arranged, yet so

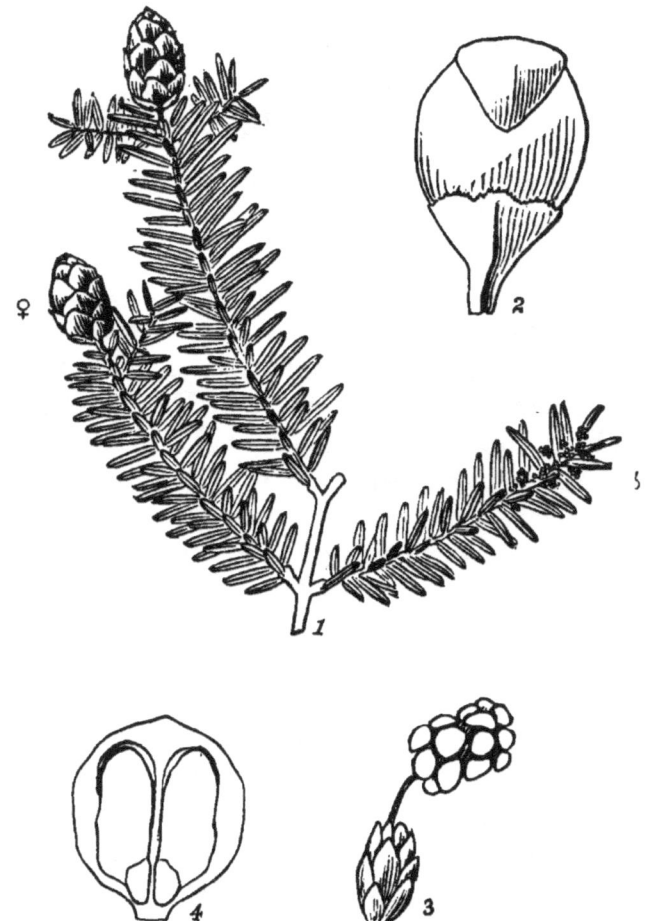

Fig. LVIII.—Abies Canadénsis: 1, a branch with fertile flowers at ♀, and sterile at ♂; 2, a scale, with its short bract; 3, a cluster of ♂ flowers (stamens); 4, a scale with its 2 perfected, winged seeds, seen from within.

inclined to a horizontal position as to appear 2-ranked on the slender spray. The ♂ aments are very small, scarcely

2″ long, each with 10-20 anthers. The pollen grains are single. The ♀ aments are terminal, ovoid, 3″ long, composed of imbricated green scales (carpels). The fruit is an oblong brown cone three-fourths of an inch long, pendant on the ends of the slender branchlets. The scales are about 20, rounded, 2-seeded. The seeds are winged, *naked* as in the Pines. This is *Abies Canadénsis*.

The Order.—From these examples the student will apprehend the nature of the CONIFERÆ (Conifers) or Cone-bearers.

>Trees and shrubs with resinous juice.
>Leaves evergreen, awl-shaped or needle-shaped.
>Flowers in aments, monœcious, without calyx or corolla.
>Ovary an open scale 2-ovuled, with no stigma.
>Seeds with pericarp, truly naked.

Classification.—With their wood growing by external layers and the embryo of 2 or more cotyledons, the Coniferæ are Exogens. But they differ from other Exogens in having *no stigma, and open carpels never inclosing the naked seeds.* Hence the division of the Exogens into two classes—the *Gymnosperms* (*gymnos*, naked, *sperma*, seeds), including the Coniferæ, and the *Angiosperms* (*angios*, a vessel, and *sperma*), including all other Exogens.

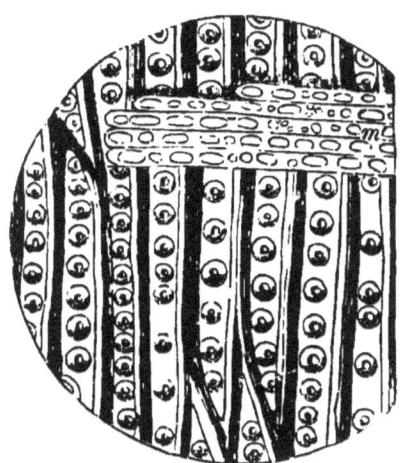

5, Pitted wood-cells of Pine greatly magnified; *m*, medullary rays.

The Conifers. — Here are associated 20 genera and 100 species, "sons of the forest and forest kings, gigantic in size, noble in aspect, robust in constitution." They inhabit all climates, but are most abundant in the North Temperate Zone. Timber and turpentine are their special products.*

The Douglass Fir (*Abies Douglásii*) of Oregon, and the Red wood (*Sequoia sempérvirens*) of California, are frequently 12 feet in diameter and 200 feet high. The Lambert Pine (*P. Lambertiána*) of California, a tree of faultless symmetry, is often 12 feet in diameter and 300 feet

View in Calaveras.

high. But over all towers the Giant Cedar of the Sierras (*Sequoia gigántea*). One grove in Calaveras County contains 90 so-called "Big Trees," measuring from 20 to 36 feet in diameter and 350 in altitude! †

---

* The wood of the Pines, Cedars, and of the Conifers generally, is remarkably distinguished by rows of circular disks which under the microscope appear like pearls bedecking each wood-cell. This form, called *pitted tissue*, has often been detected in the fossils of bituminous coal, thus revealing the origin of that useful mineral.

† Such is the perfect symmetry of these gigantic trees that the spectator finds it difficult to realize their enormous proportions. "If," says Whitney, "one could be

THE HEMLOCK.

ORGAN.	Life, Habit, Number, Place, Kind, Construction, Form, Placentation, Size, Qualities, Appendages.
Plant, L.H.S.Q.	A tree of many years growth, 50 feet high, evergreen.
Root, L.K.	Not observed.
Stem, L.H.K.F.	An erect, short, cylindrical trunk, excurrent, with many [branches.
Leaves, L.P.C.F.S.Q.	☉, spiral, acerous, sharp, subsessile, sub-4-sided, 7″.
Inflorescence, P.K.A.	In cone-shaped aments, axillary and terminal.
Flower, N.C.	Monœcious, naked.
Calyx, F.Q.	None.
Sepals, L.N.P.F.	♂ The sterile flowers in small, ovoid, red aments, axil- [lary and terminal.
Corolla, F.Q.	None.
Petals, L.N.P.F.	♀ The fertile flowers in a cylindrical, terminal ament, [with green scales.
Stamens, N.P.C.	Numerous, crowded, with short filaments.
Anther, D.C.F.	2-celled, opening lengthwise, pollen grains triple.
Style, N.C.F.	None.
Stigma, N.F.	None.
Ovary, C.F.Pn.	Carpellary scales rounded, open, thin-edged, spirally [imbricated, subtending 2 erect ovules.
Fruit, N.D.K.F.Q.	A cylindric, pendent cone, cinnamon-colored, 6′ long.
Seed, N.C.F.Q.A.	2, orthotropous, flattish, light-brown, winged.

LOCALITY.—*In Central Park, New York.* (Date) *April, 1878.*
CLASSIFICATION.—GYMNOSPERMOUS EXOGENS.
    ORDER.—CONIFERÆ, THE CONIFERS.
        NAME.—Latin, **Abies excelsa.**
            —English, *Norway Spruce.*
    REMARKS.—*Tree pyramidal in its form.*

transported to Washington and placed beside the Capitol, its summit towering far above the statue which surmounts the dome of the noble structure, the effect would be overwhelming." Various estimates have been made of the age of the Big Trees of the Calaveras Grove, and it has been poetically asserted that they were in their prime when Noah built the Ark, and may have been "contemporary with the creation of Man." The geologists of the California survey fixed the age of one tree that was cut down at 1300 years. (We counted 1362 layers.) Six feet from the ground it was 23 feet in diameter inside the bark (that being about 15 inches thick),

Timber of excellent quality is afforded by all \these species. That of the Redwood, as well as most of the Cedars, is almost indestructible. Red Cedar (*Juniperus Virginiana*) is used in the manufacture of lead pencils. The Temple of Solomon was built of the Cedars of Lebanon (*Cedrus Libāni*). The Southern Pine is heavy and fragrant with resin, affording excellent timber for floors. The Norfolk Island Pine (*Eutassa excelsa*) is celebrated for its timber and for its stately beauty.

Turpentine is distilled from the pitch which flows from the Southern Pine; resin is the residuum after distillation. Burgundy pitch is obtained from *P. sylvestris* of Europe. Canada balsam flows from the "blisters" in the bark of our beautiful *Abies balsāmea*.

Tannic acid abounds in the bark of the Hemlock; hence it is, like the Oak, extensively used in tanning leather.

The Yew tree (*Taxus*) figures in history as the favorite wood for making bows, once the formidable weapon of the English yeoman. Our Yew is a straggling shrub, never attaining the dimensions of a tree.

LIX. THE PALMETTO.*

Description.—In the forests that skirt the sandy coasts of the Southern States, the renowned Palmetto reigns. It is a tree arising 25 to 40 feet, with trunk erect, simple, 10–20' in diameter, all developed from one terminal bud. From this bud, in early Spring, a new set of leaves is annually produced above the old before they fall. Hence the tree is evergreen.

Analysis.—The *Stem* of the Palmetto exhibits new features especially worthy of study. Outside it is rugged in aspect, especially above, where it is beset with the split bases of former leaf-stalks. The trunks of other forest trees are largest at the base, diminishing upward. Not so with the Palmetto. Its trunk either continues of uniform size, or perceptibly enlarges toward the summit, there attaining its

---

* The Palmetto is the emblem of South Carolina. The massive terminal bud, consisting of numerous undeveloped leaves, is much prized as a vegetable, whence

224   THE PALMETTO.

largest diameter. For these and other peculiarities, the stem of the Palm is called a *caudex*.

Fig. LIX.—Sabal Palmetto. From a photograph. Scene in Florida, showing Fort Matanzas in the distance.

The *internal structure*, as seen in sections, may now be compared with that of Apple tree, Elm, Pine, and other Exogenous trees (p. 109). The contrast is great.

---

the tree is called the Cabbage Palmetto. To secure the bud, however, the entire tree has to be sacrificed. The wood does not splinter easily, and, on that account, was employed in building Fort Moultrie in the Revolutionary War (*Barnes's History of the United States*, p. 170). Blocks from the softer portions of the trunk are used in the South as a substitute for scrubbing brushes. The larger leaves serve for thatching and are woven into baskets or mats, while the smaller are made into hats and bonnets.

In Exogens, the bark, wood and pith are clearly defined. In Palmetto, all these are commingled; no separable bark, no woody layers, no medullary rays. The wood exists in threads or fibers extending lengthwise, traceable from the bases of the petioles down through the soft pith or *cellular tissue*, at length turning outward and ending in or at the surface where the bark should be. The composition of these wood-fibers or bundles may be understood from the cut, which represents the shaving of a Rattan magnified 100 diameters. In a single fiber there are annular cells, spiral vessels, dotted ducts, and wood-cells, all lying in the cellular tissue, *a, a*.*

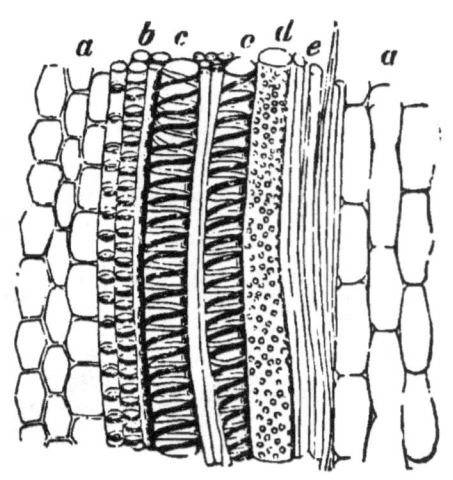

The *Leaves* are comparatively few and immense,† 7–12 feet in length, including the smooth, channeled petiole.

Various kinds of vessels in a wood-fiber of Bamboo or Rattan: *a*, cells of parenchyma; *b*, annular cells; *c*, spiral vessels; *d*, porous duct; *e*, wood-cells.

The blade is typically fan-shaped (*flabelliform*), with the border palmately cleft into many segments, in vernation *plicate*, and parallel-veined.

---

* Woody stems, whether *exogenous* or *endogenous*, are chiefly composed of the 5 classes of cells exhibited in the cut. The difference lies in their arrangement. The study of the vegetable cell, in all its varieties, is of great interest and importance, but belongs to a higher department of Botany than is admissible in this work. See *Physiological Botany*, in the Class Book, p. 130.

† Much has been written of the beauty of the Tropical Palm, decorated with its waving crown. But the Eastern traveler finds a forest of Date Palms, on the banks of the Nile, far less imposing than our own groves of Oak, Birch and Maple. Below is only a vista of naked, monotonous columns, and above a scanty foliage through which the rays of the sun pour in undiminished intensity.

The *Flowers,* open in June, are perfect, sessile, on a long branching spadix with bracts or a double spathe at each joint. They have a double perianth, of 3 sepals and 3 petals, 6 stamens, and a triple pistil which in fruit becomes a single 1-seeded, round drupe, like a date.

The Name.— *Sabal Palmetto* is the only Palm in the United States which attains the dimensions of a tree. Two other species of this genus, the Dwarf and the Saw Palmetto, form dense thickets in the wilds of the South. They are mere shrubs, with caudex prostrate or creeping. The Blue Palmetto, with caudex 2 or 3 feet long, erect or prostrate, has *polygamous* [some ♂, some ♀, and some (perfect) ☿] yellowish flowers, and is assigned to another genus—*Chamærops Hystrix.* We have no other Palms.

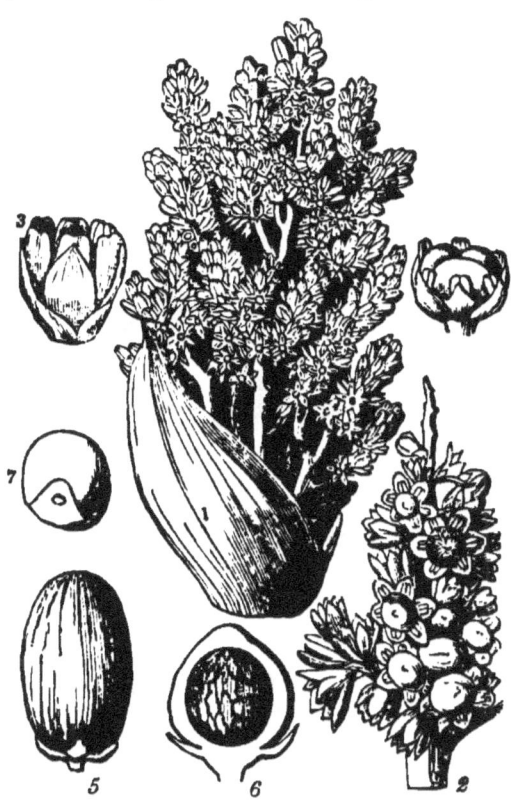

1, Inflorescence of Chamærops humilis, in its spathe; 2, a branch of the same with the fruit ripening; 3, a sterile ♂ flower; 4, a fertile ♀ flower; 5, a ripe fruit; 6, a section of another variety, showing the seed; 7, section of seed showing the embryo. (From *Lindley.*)

The *Cocoanut* is a fruit of similar construction, and its seed is, perhaps, the largest of all seeds. Let it be analyzed in this connection. Like other drupes, this also has two coats, the outer of loose, woody, brown fibers, the inner a shell of bone. At the apex of the shell are 3 apertures—the scars of the stigmas. Within the shell is only 1 cell and 1 seed, although the ovary was 3-celled and 3-ovuled. The cut (11) shows a section of the seed—the white, fibrous, oily albumen with a cavity which contained the milk—and at *e*, the embryo, 1-cotyledoned, in a separate, smaller cavity;

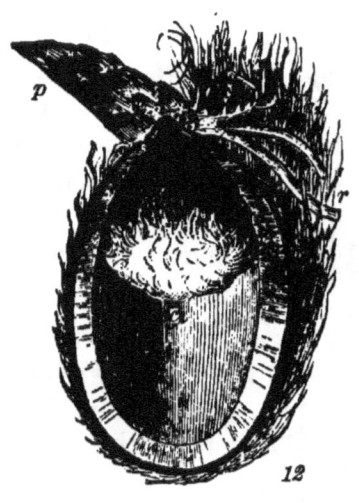

11, section of the seed of a Cocoa-nut; *e*, the embryo; 12, Cocoa-nut germinating.

(12) shows its germination —the growing plumule *p*, the growing radicle *r*, and the enlarged cotyledon *c*, partly filling the cavity.

Classification (ordinal).—The order PALMACEÆ is estimated at seventy-three genera and four hundred species. Nearly all are natives of the Torrid Zone in both hemispheres. The Palms rank among the noblest of the Vegetable Kingdom, whether we regard their towering stems, their magnificent leaves, their numberless flowers, or their valuable products. The trunks of some attain the height of one hundred and eighty and a diameter of five feet.

Cálamus Rudéntum, of the Malaccas, grows in the form of a cable five hundred feet in length dangling from trees to which it clings by the hooks on the end of its leaf-stalks. The Date Palm develops two hundred thousand flowers on a single spadix. Among its products are starch, sugar, oil, wax, edible fruits, material for clothing, building, paper-making, and fermented liquors.

The Cocoanut Palm (*Cocos nucifera*) is perhaps put to a greater number of uses than any other tree in the world. Its wood, called porcupine wood, takes a beautiful polish. The fibers of the outer covering of the fruit are very durable, and are manufactured into cordage, matting, door-mats, scrubbing brushes, etc. The inner shell is made into water-dippers. The milk contained in the cavity of the albumen is a beverage as delicious as the albumen is wholesome. The nuts by pressure yield the rich oil of cocoa. From the wounded spadix flows a sweet sap, a quart a day for several months. If boiled, it produces sugar. When fermented, it is called palm-wine or toddy, and when distilled, the vile liquor, arrack. The leaves furnish thatch for dwellings and material for fences, hats, baskets and buckets, and even paper which is written upon with a style. Potash in abundance is obtained from the ashes.*

The Sago Palm (*Sagus Rúmphii*) of Malacca, and other Palms, afford the starchy food called Sago. This exists in the cellular tissue of the stem, whence it is washed out and granulated. A single tree will yield six hundred to eight hundred pounds.

The Date Palm (*Phœnix dactylifera*), of Northern Africa, supplies that sweet and delicious fruit, the date, which furnishes the tribes of Fezzan and Barbary nine-tenths of their living.

---

* "After an abundant repast, the traveller inquires of his Indian host, Who in this desert country furnishes you with all these luxuries? My Cocoa-nut tree, is the reply. The acidulous drink tasted on your arrival was drawn from the fruit before it was ripe. This kernel, so delicate in flavor, is the ripe fruit. This milk which you find so agreeable is drawn from the nut. This cabbage, so delicate in flavor, is the top of the Cocoa-nut tree—a costly dish, however, for it takes the life of the tree. This wine is Palm-wine, drawn from the thick leaves sheathing the flowers. Exposed to the sun, it becomes vinegar; and by distillation we get this good brandy which you have tasted. This juice also supplies the sugar for these sweetmeats. Out of the shell of the nut we make these vessels and utensils. Nor is this all. This habitation itself I owe to these trees. With their wood my cabin is constructed, and with their plaited leaves it is thatched. Made into an umbrella I walk under their shade. My clothing is spun from their leaf-fibers, and these mats so generally useful are made from them also. This sifter was ready-made to my hands in the axils of the leaf-stalks. With these same leaves we make sails for our ships, and for caulking them nothing is so good as the fibers which envelop the nut. Of this, too, we make all sorts of strings, cables, and cordage. Finally, the delicate oil which has seasoned many of our dishes and that which burns in my lamp, is expressed from the fresh, ripe kernel."

Rotang (*Cálamus Rudéntum*, etc.), growing slender and to great length, affords rattan for canes, chair-bottoms, etc.

Ivory Palm (*Phytélephas*) of the Magdalena River region, contains in its seeds a compact albumen—the vegetable ivory of commerce.

The bruised fruit of *Elais Guineënsis* yields the palm-oil which is imported from Africa in immense quantities, for soapmaking and other uses.

Classification (provincial).—In a higher sense the Order of the Palms represents the grand province of the Endogens, as the Roseworts, the Mastworts, etc., represent the Exogens. These two grand divisions constitute the subkingdom Phenogamia or Flowering Plants. They are severally marked by the following five characters, which we place in contrast and arrange in the *descending* order of their value, that is, their constancy:

THE EXOGENS.*	THE ENDOGENS.*
Embryo with 2 or more cotyledons.	Embryo with one cotyledon.
Radicle forming an axial root.	Radicle never forming axial root.
Stem growing by accretions external to the wood.	Stem growing by scattered internal wood-fibers or bundles.
Flowers 4 or 5 (rarely 3) parted.	Flowers almost always 3-parted.
Leaves very generally net-veined.	Lvs. very generally parallel-veined.

LX. JACK-IN-THE-PULPIT.

Description.—The voice of this little declaimer is heard, if at all, in the flowery month of May, throughout the damp old woods. The plant stands about a cubit in height, with club and canopy and lurid coloring—a form so singular that to be seen is to be remembered.

---

* To apply the above classification, let the student now determine the Province to which the foregoing orders—any or all of them—belong. And generally, it will hereafter be his pleasure to view all plants in the light of these distinctions.

FIG. LX.—Arisæma triphyllum : *b*, spadix with ♂ and ♀ flowers ; *c*, flowers enlarged ; *d*, spadix with ♂ flowers ; *e*, with ♀ flowers ; *h*, berries ripe ; *g*, berry dissected.

Analysis.—The *Stem*. The base of the plant is enlarged into a kind of bulb, which being solid (not made up of scales) is called a *corm*. The shape of this bulb has given to it the common name of Indian Turnip. It consists of starchy matter pervaded by a fluid fiercely acrid to the taste, and well meriting the name "Dragon-root."* Encircling the edge of the corm is a row of fibrous roots. Evidently the corm is the stem; there is no other. A scape and 2 leaf-stalks arise from the corm, the former inclosed below by the sheathing bases of the stalks. The leaves are 2, trifoliolate. The leaflets are often as large as 4' by 6', ovate, inclining to *rhombic*, entire, acuminate. The venation is pinnate and netted, with marginal veins.

Inflorescence. The scape varies in height from 6' to 2 feet, but is never so high as the leaves. At the top is a club-shaped inflorescence called *spadix*, protected by a large bract named *spathe*. The spathe is convolute below and inflected above, colored with stripes of purple within. The spadix is naked and brown above, bearing the flowers below.

The *Flowers* are *monœcious* (♂, *monos*, one, *oikos*, house) *b, c*, sometimes *diœcious* (*dis*, two, *oikoi*, houses), *d, e*. When together, the ♂ are above the ♀, and consist of 4 or more sessile anthers opening at the top. The ♀ fertile flower is merely a 1-celled ovary with flat stigma and 2 or more ovules erect from the bottom of the cell (*g*).

A section of the seed (*g*) shows a straight embryo in the midst of fleshy albumen with only 1 cotyledon. The fruit is a mass of scarlet, several-seeded berries.

The Name of this plant is *Arisœma triphyllum*—Ari-

---

* The starch in many species of this plant is used as food. In the days of Queen Elizabeth it furnished the stiffening for the enormous lawn ruffs then worn by gentlemen and gentlewomen. These became so large that it is said the Queen placed a guard at the city gates to cut down any ruffs that were over a yard wide. They needed a very strong starch, such as was made from this root; though it was, says the old herbalist, "most hurtfull to the hands of the laundresse, for it chappeth, blistereth and maketh the handes rough and withall smarting."

sæma being an alteration of Arum, its former name; triphyllum, the same as trifoliate, or trifoliolate.

Arisæma Dracóntium, Green Dragon, another species growing in marshes, has its one leaf divided into about 9 leaflets, and its spadix very long-pointed.

7, Oróntium Americànum ; *s*, the spadix destitute of a spathe ; 8, Calla palústris ; *b*, a spathe and spadix ; 9, a flower with 6 stamens and an ovary ; 10, cross-section of a berry, showing 6 cells.

The Golden Club (*Oróntium*) growing in rocky rills, may be examined in connection with the Arisæmas. Its yellow spadix has no spathe and is covered above with perfect flowers. Also our native Calla (*C. palustris*) growing in swamps, whose short spadix is covered with perfect flowers and invested with a white spathe.

The favorite House Calla (*Richárdia Africàna*) is a native of S. Africa. Here, also, the flowers cover the whole spadix, the ♂ above and the ♀ below. The leaves of this and of the two preceding are decidedly parallel-veined.

Classification.—The order ARACEÆ—the Aroids—includes the above and many other genera. We may briefly define the order as follows :

Herbs pungent and acrid, with rhizomes or corms.
Leaves often net-veined, generally parallel-veined.
Flowers small, crowded on a spadix.
Ovary free, with a sessile stigma.
Embryo with one cotyledon.

The Aroids are chiefly tropical, numbering 46 genera and 240 species. They are generally acrid, and some are dangerously poisonous.

The Dumb Cane (*Diffenbáchia*) of the W. Indies is so called because, if tasted, it causes the tongue to swell and fill the mouth.

Sweet Flag (*Acorus*) grows in cold streams of the Northern States. Its long, thick rhizomes are sought for their warm, pungent, aromatic taste.

Caládium and *Colocásia* are cultivated for their large, ornamental leaves; also for their tuberous, edible roots.

Scientific Terms.—Diœcious. Monœcious. Spadix. Spathe.

LXI. THE SHOWY ORCHIS.

Description.—With eager longing and patient search the botanist expects the Rose-tinted Orchis in the late days of May, when Spring is fading into Summer. It belongs, with Lady's Slipper, to a high-toned, fastidious race, very choice of its soil in old rich woods, here and there, and soon retreating when its haunt is discovered. It will be promptly recognized by its two obovate, shining leaves, 4-angled scape, and several rose-colored flowers.

Analysis.—The root, bract, leaves, and scape, we leave to the discrimination of the student. Let him note every point of form or structure whereby the species may be distinguished from others.

The *Flower* is constructed after a pattern quite new and extraordinary. In general aspect it seems bilabiate. Beneath is seen the inferior (*adherent*), twisted ovary. A careful analysis will show the perianth composed of 3 sepals in an outer whorl and 3 petals in an inner one. The lower petal

Fig. LXI.—Orchis spectábilis: 2, a flower; *l*, the lip; *s*, the spur; *o*, the twisted ovary; 3, the column (enlarged); *b*, the place of the sticky glands at the ends of the stalks of the pollinia, which are seen partly extracted from the anther cells; *a*, the stigmatic surface; 4, a pollinium (pollen mass) adhering to the finger, at first erect, soon declined as when attached to a moth's head, in order to be thrust in the face of the stigma.

is the lower lip, and it is at the base produced backward into a slender spur—the nectary—seen under the ovary. The two upper petals are somewhat united, covering the stamens like a hood. The 3 sepals are also ascending and converging with them—all rose-purple, forming a vaulted upper-lip.*

Instead of stamens and pistils, there is an oval, concave mass called the *column*—a stamen and pistil combined. In it are 2 anther-cells, and a broad stigma-surface between them. Each cell contains a club-shaped mass of granular pollen, erect on a stipe attached to a sticky gland on the stigma. These pollen masses are the *pollinia*, and such flowers are called *gynandrous*.†

5, Calopògon pulchéllus—lip on the upper side (*l*), column on the lower.

Fertilization. In the Orchis it becomes an interesting question how the pollinia shut up in the 2 cells *c c* can be brought into contact with the stigma at *a*? Repeated obser-

---

* Strictly speaking, the lower-lip; for by the twisting of the ovary half-a-turn the whole flower is inverted. In the elegant Grass Pink (*Calopogon*) the ovary is not twisted; and the lip proper, the lip consisting of one petal, is on the upper side of the flower.

† Every part of the Orchis seems purposely shaped to perform some special work in its economy. The upper portion protects the delicate pollen-masses; the hinder-part is prolonged into a tube which does not yield nectar until it is gnawed; and the lower portion is a tempting and convenient alighting stage for insects. Even the pollen-masses are specialized in a marvellous degree, and are usually associated with mechanic contrivances intended for adherence to the proboscea and bodies of butter-flies and moths. Their perfumes are as various as their shapes, and even the honey seems to have a variety of flavor which makes it more sought for by some insects than by others. Had the Orchids been rational beings fully aware of the laws of biology, chemistry and mechanics, they could not have adapted themselves to their surroundings more perfectly.—(*Taylor*.)

vation has shown that the agent is a butterfly with a proboscis long enough to fathom the nectary.* The mouth of the nectary opens just beneath the stigma, and close by the two sticky glands already named. The lip is the platform on which the insect alights. Thrusting its proboscis into the opening in order to reach the nectar, it comes into contact with the glands, which adhere to its head so that in retreating it drags the pollinia from their cells. You may do the same with your finger (4) and observe that in a few seconds the pollinium bends downward on its stalk (*d*). Thus it comes into the exact position to be dashed against the stigma below the cells of the next flower the insect visits. In this way, crossing is almost inevitably secured.

The Name.—Of the genus *Orchis* we have more than 20 native species. All may be recognized by the agreement of the flowers to the above description. The species represented in Fig. LXI, *O. spectábilis* (Showy Orchis), is the earliest in flower. Specifically it differs from the others in its stigma, which has the sticky glands near together and enclosed, while in the others they are separated and naked; and the few rose-purple flowers are on a square scape not taller than the 2 obovate, obtuse, parallel-veined leaves.†

Scientific Terms.—Column. Gynandrous.

---

* A Madagascar Orchis has a nectary nearly a foot long. Darwin inferred from this fact that huge moths would be found on that island with proboscees sufficiently long to thrust down this lengthy tube. No moth in any part of the world was known to possess such a proboscis. Since then Müller has found a species in Brazil with a trunk that even when dried is ten or eleven inches long.

† All the European Orchids and some of our own (*Adam-and-Eve*, e. g.) have roots with two lobes—one hard and vigorous, the other withered and decaying. The former is an offshoot from the latter, and has, on its own opposite side, a bud that will the next year expand and send up a stem, the new plump bulb withering in its turn. In consequence of this mode of growth, the position of the plant changes about half an inch every year, and so

"The Orchis takes
Its annual step across the earth,"

in time becoming quite far removed from its original position.

LXII. OTHER ORCHIDS.*

Description.—Among the flowers of June, the practiced collector, acquainted with *O. spectábilis*, will recognize yet other kinds of Orchis, or of its cousins. In old woods abounding in Hemlock and Pine, let him expect the Great Round-leaved Orchis (*O. orbiculàta*). It will be known by its 2 large (5′–8′), rounded, polished, parallel-veined leaves lying flat on the ground, and its tall (2 ft.), bracted scape bearing a raceme of straggling greenish-white flowers.

Fig. LXII.—A flower of Orchis Psychòdes : *o*, the twisted ovary ; *s*, the spur.

Analysis.—The student will mark the form of the column, so different from that of *O. spectabilis*. Here the sticky glands are naked, disk-form, and widely apart, as are also the pollinia which they support. Note also the long strap-shaped lip, the roundish upper sepal, and the slender spur nearly 2′ long. No insects but the largest Sphynx Moths

---

* In the Orchids are seen the highest evidence of the mutual relations of flowers and insects. In numerous species special adaptations are carried so far that while self-fertilization is impossible, the service of crossing is limited to a single species of insect only. Thus Nature here emphasizes the principle of cross-fertilization ; and experience has shown that plants raised from seed produced by flowers fertilized by pollen from another flower, are stronger, usually taller, bear more flowers and produce more seeds than those grown from the seed of plants fertilized by their own pollen. "Nature seems everywhere to have forbidden the banns of intermarriage, and her decree is carried out whenever possible, from mosses to men." There have been cases in the history of some tribes where intermarriage only could save the race from extinction. Similarly, among plants, there are some flowers which have to adopt self-fertilization as a rule, or as a last resort ; but the pre-eminent law is against it even more in plants than in animals, enforced by the very structure of the flowers. It is hardly too much to assert that every species of flowering plant has its peculiar modification to realize this end. It is as if plants themselves were conscious of the importance of this principle, and so adopt some device to carry it out.

have a proboscis long enough to fathom the depth of this nectary, suck its nectar and extricate its pollinia.

Purple-fringed Orchis. In June and July, look in wet grassy meadows for these handsome Orchids, known by their erect, tall stems, beset with lance-shaped leaves below, bracts above, and a terminal plume-like raceme of roseate purple-fringed flowers. The twisted ovary, long slender spur, the lip 3-parted and fringed, the 2 other petals either notched or fringed, are their striking features. Then the column, protruding forward and apart, the 2 button-shaped glands of the cooped pollinia, present altogether an appearance extravagant and grotesque.

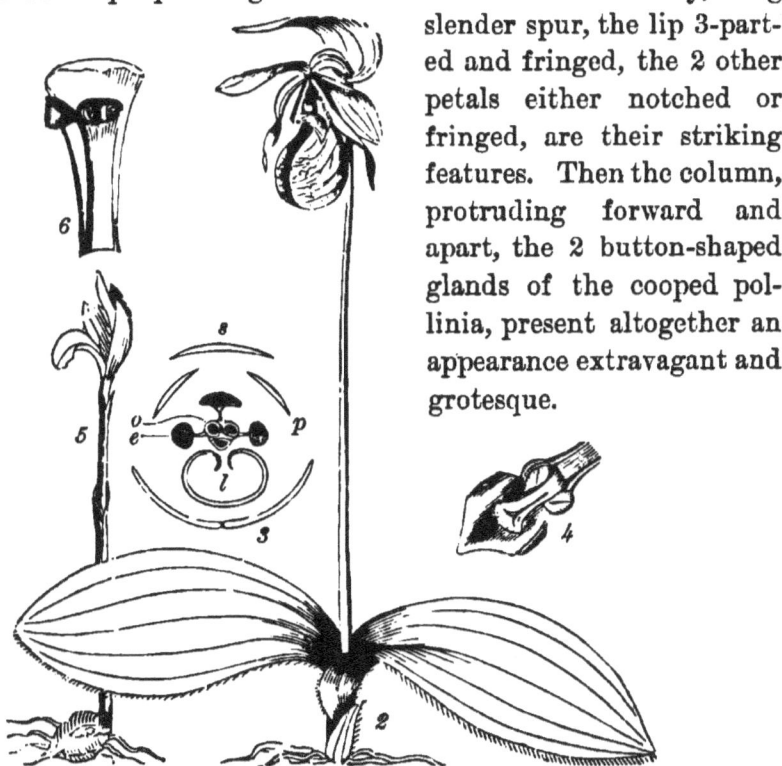

2, Cypripèdium acaùle ; 3, plan of the flower ; *s* (outer circle), 3 sepals, the 2 lower united, *p*, the petals, one of them (*l*) the lip, *e*, the stamens, *o*, the ovary ; 4, the column, seen from beneath, showing the 3 stamens, 1 leaf-like, and stigma ; 5, Arethùsa bulbòsa ; 6, the column, with lid-like anther opened by the bee.

Fig. LXII is an enlarged view of a flower of *O. Psychòdes* (*psychè*, a butterfly, *eidos*, appearance). With a lip more deeply fringed, and the two other petals, *p p*, also fringed, the cut would nearly resemble the flower of Dr. Bigelow's *O. grandiflòra,* so common in the meadows of New England. Again, with the lip less deeply fringed, the terminal segment split, the lateral segments squarely clipped, and the whole flower violet-purple, we should have a flower of *O. peramœna,* a splendid plant of the meadows West and South.

Lady's-slipper. Several kinds of this interesting genus (*Cypripèdium*) may be detected in their sylvan retreats, and analyzed in this connection. Three distinctive marks will be noted. The column has 3 stamens, 2 with anthers and 1 petal-like, with no anther; the lip is an inflated sack; the 2 lower sepals are united to near the apex.

Classification.—These few instances suffice to introduce the great and marvelous order of the ORCHIDACEÆ—the Orchids—containing probably 400 genera and 3000 species, known by the following marks:

> Herbs with parallel-veined leaves.
> Flowers irregular, 3-parted, with a lip.
> Perianth adherent to the ovary.
> Stamens 1 or 2, gynandrous.
> Pollen cohering in masses—pollinia.
> Ovary 1-celled, with innumerable ovules.
> Fertilization effected only by insect agency.

The Orchids grow in all countries, but are most abundant in the hot damp regions within the Tropics. There they thrive in countless thousands as air-plants (*épiphytes*) independent of the soil, clinging to the trunks and branches of trees, and to naked rocks, drawing their nourishment from the air alone, displaying curious and grotesque forms of floral beauty in endless variety. Their mimicry of insects, birds [*] and reptiles, is often very striking, and also significant in view of their dependence on insects for their very existence. Are the insects themselves deceived and enticed by these animal forms and appearances?

The products of the Orchids useful to mankind are very few. The

---

[*] Thus in the Holy Spirit Plant (*Peristèria elàta*) of Central America, the corolla is of alabaster whiteness, and the column within is an almost perfect likeness of a dove with outspread wings, as artists are wont to paint the Holy Spirit. No wonder that among the ignorant natives it becomes an object of superstitious reverence,

fragrant Vanilla used in confectionery is obtained from the dried fruit of *Vanilla planiflora*, and other species, of Mexico.*

ORGAN.	*L*ife, *H*abit, *N*umber, *P*lace, *D*ehiscence, *K*ind, *C*onstruction, *F*orm, *Pl*acentation, *S*ize, *Q*ualities, *A*ppendages.
Plant, L.H.S.Q.	♃, *terrestrial, acaulescent, 1 foot high, downy.*
Root, L.K.	♃, *inaxial, of many tufted fibers.*
Stem, L.H.K.F..	*Crown subterranean, undeveloped.*
Leaves, L.P.C.F.S.Q.	☉, *2, radical, parallel-veined, plaited, ovate, 5'.*
Inflorescence, P.K.A.	*Scape with a bract and a flower at the top.*
Flower, N.C.	*1, irregular, symmetrical, 3-parted, perfect.*
Calyx, F.Q.	*Open, adherent, greenish.*
Sepals, L.N.P.F.	*Deciduous, 3, the 2 lower united, lance-oblong.*
Corolla, F.Q.	*Very irregular, the lip rose-purple.*
Petals, L.N.P.F.	*2 petals linear, lip saccate, obovoid, large.*
Stamens, N.P.C.	*3, joined to the pistil, 1 a rhombic leaf.*
Anther, D.C.F.	*2, fertile, 2-celled, with granular pollen.*
Style, N.C.F.	*Short, under the sterile stamen.*
Stigma, N.F.	*Terminal, obscurely 3-lobed, roughish.*
Ovary, C.F.Pn.	*Inferior, curved, oblong, 1-celled.*
Fruit, N.D.K.F.Q.	*A 3-angled, 3-valved capsule, 1-celled.*
Seed, N.C.F.Q.A.	*Very numerous and minute.*

LOCALITY.—*Rocky woods, Stamford, Conn.* (Date), *May, 1865.*
CLASSIFICATION.—FLOWERING ENDOGENS.
—Order, ORCHIDACEÆ, THE ORCHIDS.
NAME.—Latin, Cypripedium acaule.
English, *The Stemless Lady's-slipper.*
REMARKS.—*The lip is slipper-shaped, near 2' long.*

* Attempts made to grow this plant in the East Indies failed, since, though the plant blossomed abundantly, it failed to fruit, owing to the absence of the insect that in its native haunts is its pollen-bearer. On that account artificial fertilization was adopted,

LXIII. IRIS, OR BLUE FLAG.*

Description.—The Blue Flag is everywhere associated with swimming bogs, bull-rushes, and frogs. We look for the large blue flowers in June. Several new features, and new combinations of old ones, here await the student's pleasure.

Analysis.—The base of the stem will remind him of the rhizome of the Bloodroot. The leaves are decidedly parallel-veined, and arranged alternately, in 2 ranks. As to form and position, they are *ensiform* (*ensis*, a sword); that is, linear in outline, and *vertical*, or with the edges turned upward and downward. The vernation of the leaves, as seen in a cross-section of an early shoot (2), is *equitant* (i. e. riding astride).

The *Flower.* Note the inflorescence, and bracts; the convolute æstivation of the calyx and corolla as seen in a cross-section of the bud; and the position of the floral organs exhibited in a vertical section (3), the latter showing the adhesion of the perianth to the (inferior) ovary. The attitude and colors of the sepals and petals are remarkable; the former reflexed, the latter erect or even incurved over the central organs. The 3 stigmas would hardly be known were it not for their position, so much do they resemble petals. These are properly winged styles, only the tip of the upper surface being stigmatic. At the base, they

---

* The Iris has an historic interest. Several species of the flower have borne the name flower-de-luce or fleur-de-lis, from the French Fleur-de-Louis, as during the Crusades that king adopted it as the emblem of his shield, and strewed it on the mantle of his son at the coronation in the cathedral at Rheims (*Brief History of France*, p. 46). After the battle of Crecy, it was united with the arms of England, but afterward gave place to the Shamrock of Ireland. It is still the Lily of France. The ancients regarded the Iris as the emblem of eloquence or power. It was placed on the brow of the Sphinx, and the kings of Babylon bore it on their scepter.

Fig. LXIII.—Iris versicolor: 2, section showing the vernation; 3, vertical section of the flower; 4, the flower displayed, excepting 5, the pistil and 3 stigmas.

unite with one another and with the perianth, and so continue down to the triple adherent ovary.

The *Fruit*, when mature, is a dry, oblong, obscurely 3-cornered capsule, 3-celled and 3-valved. The seeds are numerous. A dissection shows them to contain a minute monocotyledonous embryo in a large albumen.

Fertilization.—Half concealed beneath the arching stigmas we find the 3 stamens, with the anthers *extrorse*, that is, opening and discharging their pollen outward—averse from the stigma. This suggests the inquiry, How can the pollen from the anthers ever reach the stigmas? Instead of facilitating, special care seems to have been taken to guard against it; the anther and stigma placed back to back, the former beneath and shedding its pollen downward. It is clear that self-fertilization is impossible. In the cavity at the bottom of the flower is a drop of nectar. First, the attention of the foraging bee is caught by the gay colors of the flower; then he is drawn by the nectar. Alighting on a spreading sepal he brushes the anther both coming and going, catches on his head and back more or less of the pollen, which will thus be carried and scattered either on the stigma of the same flower, or of the flower next visited, where also he is again dusted. The result is an endless interchange of pollen, with a greater probability of *cross* than of *close* fertilization.

The Name.—*Iris versicolor* is the classical name—Iris, the fabled deity of the Rainbow; versicolor, various-colored. Other species of Iris will be found in the meadows and bogs, and still others in the gardens. Let them be compared with this and their differences noted.

LXIV. BLUE-EYED GRASS.

Description.—This is a frequent inhabitant of meadows, both lowland and upland, so much resembling the grasses in its foliage that it would seldom be detected but for the blue flowers open in June.

Analysis.—What kind of a root has it? Its life? The stem—has it any? And what its place, size, etc. ? The leaves—are they vertical as in Iris?

The *Inflorescence*—is it borne on a stem or a scape? Respecting its form, the stalk is *ancipital*—two-edged or winged. The flowers issue from the midst of several bracts, of which the 2 outer are green, the inner *scarious*. The ovary—is it free, or adherent? The perianth is 6-parted. Is there any difference between the 3 sepals and the 3 petals? All are *mucronate*, that is, tipped with a slender point, with a notch. The stamens are 3, with their filaments united below into a tube sheathing the triple style. The fruit is a globular capsule with 3 cells and about 24 roundish seeds. The seed, black and rough-coated, shows, in a section, an embryo with one cotyledon in large albumen.

The Name is *Sisyrínchium Bermudiána*—the former from the Greek, meaning a pig's snout, alluding to the form of the bracts; the latter referring to the islands whence Linnæus first received it.

The beautiful *Gladiolus* (a little sword) of the gardens, with ensiform (sword-shaped) leaves and a spike of irregular flowers, may be analyzed in connection with Iris and Sisyrinchium.

The Order of the Irids—IRIDACEÆ—is represented by these plants. It includes 55 genera and 550 species, chiefly natives of S. Africa, Europe, and the United States. They coincide in the following traits:

Fig. LXIV.—Sisyrínchium Bermudiàna: 2, vertical section of the perianth, showing the ovary (*o*, inferior), the 3 united stamens, etc.; 3, the ovary dissected; 4, cross-section of the capsule; 5, a seed dissected, showing the embryo; 6, plan of the flower of an Iris.

Herbs with equitant, 2-ranked leaves.
Perianth adherent to the ovary.
Segments in 2 sets, contorted in bud.
Stamens 3, with extrorse anthers.
Stigmas 3, opposite to the stamens.
Capsule 3-celled, 3-valved.
Seeds many, with hard albumen.

The Irids are more remarkable for beautiful and fugitive flowers than for useful products. Besides those already noted, we find here the Mexican *Tigridia*, the *Crocus, Tritonias, Watsonias*, etc.

The Aromatic Orris-Root used in all tooth-powders, with the aroma of Violets, is obtained from the Florentine Iris (*I. Florentina*).

Saffron, a well-known yellow dye as well as a useful medicine, is the dried stigmas of *Crocus sativus*.*

LXV. THE TRILLIUMS.

Description.—These plants are peculiarly interesting for the symmetry of their construction, the *ternate* division being extended in them throughout. They may be sought generally in shady forests, which they enliven with their conspicuous bloom in the month of May, rarely earlier. While the genus is widely disseminated in the country, the species are local, no one kind being within the reach of all. It is therefore expedient to direct attention to the

GENERIC CHARACTERS.—The *Stem* arises from a tuberous root-stock always *premorse*, i. e., ending abruptly as if bitten off, in consequence of the decay of the portions grown in previous years. Each plant has 3 leaves and 1 flower.

The *Leaves* are verticillate and exhibit a reticulate venation, as if the plant were an Exogen, which it is not.

---

* Saffron is referred to in the Song of Solomon as a valuable product. In ancient Greece and Rome, it was a condiment highly esteemed on the tables of the rich. Our Saxon ancestors relished jusselle—a compound of eggs, grated bread, saffron and sage; and an old English writer says, "Without saffron we cannot have well-cooked peas." Saffron is still used at the East as a perfume. As only the *stigma* of the Crocus should be used, the drug is costly, and is often adulterated with the florets of Marigold and Safflower (*Cárthamus tinctòrius*), which want the aromatic and stimulating properties of the Crocus-saffron.

ORGAN.	Life, Habit, Number, Form, Dehiscence, Kind, Construction, Form, Placentation, Size, Qualities, Appendages.
Plant, L.H.S.Q.,	♃, *herb erect, 15' high, glabrous.*
Root, L.K.,	♃, *rhizome short, thick, premorse.*
Stem, L.H.K.F.,	*herbaceous, caulis erect, simple, terete.*
Leaves, L.P.C.F.S.Q.,	*a whorl of 3, terminal, broad-rhombic, some net-veined.*
Inflorescence, P.K.A.,	*terminal, solitary, peduncle erect, or nearly so.*
Flower, N.C.,	*one, perfect, complete, regular, 3-parted.*
Calyx, F.Q.,	*wheel-shaped, green.*
Sepals, L.N.P.F.,	*persistent, 3, spreading, lance-ovate.*
Corolla, F.Q.,	*star-shaped, dark dull purple, ill-scented.*
Petals, L.N.P.F.,	*deciduous, 3, spreading, lanceolate-ovate, pointed.*
Stamens, N.P.C.,	*6, erect, included, filaments short.*
Anther, D.C.F.,	*linear, adnate, introrse, opening lengthwise.*
Style, N.C.F.,	*3, stigmatic along the inner side.*
Stigma, N.F.,	*3, long, recurved and revolute.*
Ovary, C.F.Pn.,	*3-celled, 6-angled, placentæ central.*
Fruit, N.D.K.F.Q.,	*an indehiscent, purple, ovoid berry.*
Seed, N.C.F.Q.A.,	*several, anatropous, albuminous, black.*

LOCALITY.—*Woods, Windsor, Vt.* (Date), *May, 1850.*

CLASSIFICATION.—FLOWERING ENDOGENS.

—Order, TRILLIACEÆ, THE TRILLIADS.

NAME.—Latin, **Trillium erectum.**

—English, *Bath Flower.*

REMARKS.—*Peduncle 3' long, flower nearly 2' broad.*

THE TRILLIUMS.

The *Flower* is perfect, complete, regular and alternating. The calyx is free from the ovary, consisting of 3 green sepals. The corolla contains 3 colored petals. There are 6 stamens

FIG. LXV.—Trillium erythrocárpum : *a*, diagram, representing the floral organs apart, *s*, calyx, *c*, corolla, *st*, stamens, *o*, pistils ; 3, the flower of T. erectum with perianth removed.

evidently in 2 circles, with anthers longer than the filaments. One ovary appears, superior, compounded of 3 carpels, there being 3 distinct, sessile stigmas. The fruit, ripe in July and August, is a red or purple berry, with 3 cells. There are several seeds in each cell, having the 1-cotyledoned embryo in copious albumen.

The Name, *Trillium*, given to the genus, means triple, all the parts of the plant being in threes. There are ten species, all American. *T. erythrocárpum* (red-fruited), the

Wake Robin, illustrated in the cuts, inhabits cold, damp, often mountainous woods, from Canada to Georgia. It is known by its petiolate leaves, pedunculate flower, and white, purple-veined petals.

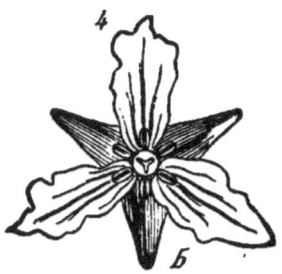

T. erectum, the Bath-flower, abounds in rich woods, especially Northward. The leaves are sessile, the flower pedunculate, and the petals dark-purple, ill-scented. A variety has white petals. The student will analyze the various species when found, and note in his Plant Record their specific characters.*

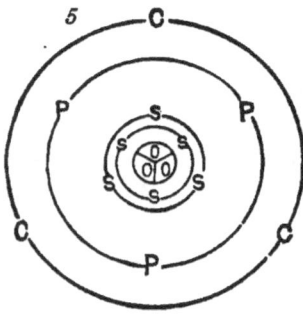

4, *b,* flower of Trillium erythrocarpum : 5, plan of the flower, *c,* the position of the sepals, *p,* of the petals, *s,* of the stamens, *o,* of the 3 united ovaries.

The Indian Cucumber (so called from its white, fleshy rhizome) is another herb of singular symmetry, frequent in the old forests of most of the States. It has a simple, slender stem, about 2 feet high, bearing a whorl of 6–8 leaves near its middle, another of 3 leaves at the top,

---

* The following analytical table, condensed from the *Botanist and Florist*, shows at a glance the distinctive features of our 8 species of Trillium :

§ Flowers sessile. Petals dark purple, erect ....Nos. 1, 2.
§ Flowers on a peduncle raised above the leaves........................(*)
　* Leaves petiolate, ovate, rounded at the base.......................Nos. 3, 4.
　* Leaves sessile, rhomboidal, nearly as broad as long................Nos. 5, 6.
§ Flowers on a peduncle deflexed beneath the leaves, white...............Nos. 7, 8.
　No. 1. *T. séssile,* L. Leaves sessile, mottled, petals sessile.
　　2. *T. recúrvum,* Beck. Leaves petiolate. Petals narrowed to a claw.
　　3. *T. nivále,* Riddell. Leaves obtuse. Petals obtuse, snow white.
　　4. *T. erythrocárpum,* Mx. Leaves acuminate. Petals pencilled with purple.
　　5. *T. grandiflórum,* Salisb. Petals obovate, 2', white, becoming roseate.
　　6. *T. eréctum,* L. Petals ovate, dark-purple, or white, 1' long.
　　7. *T. cérnuum,* L. Leaves rhomboidal. Petals spreading. Stigma distinct.
　　8. *T. stylósum,* Nutt. Leaves elliptical. Petals recurved. Stigmas half-united.

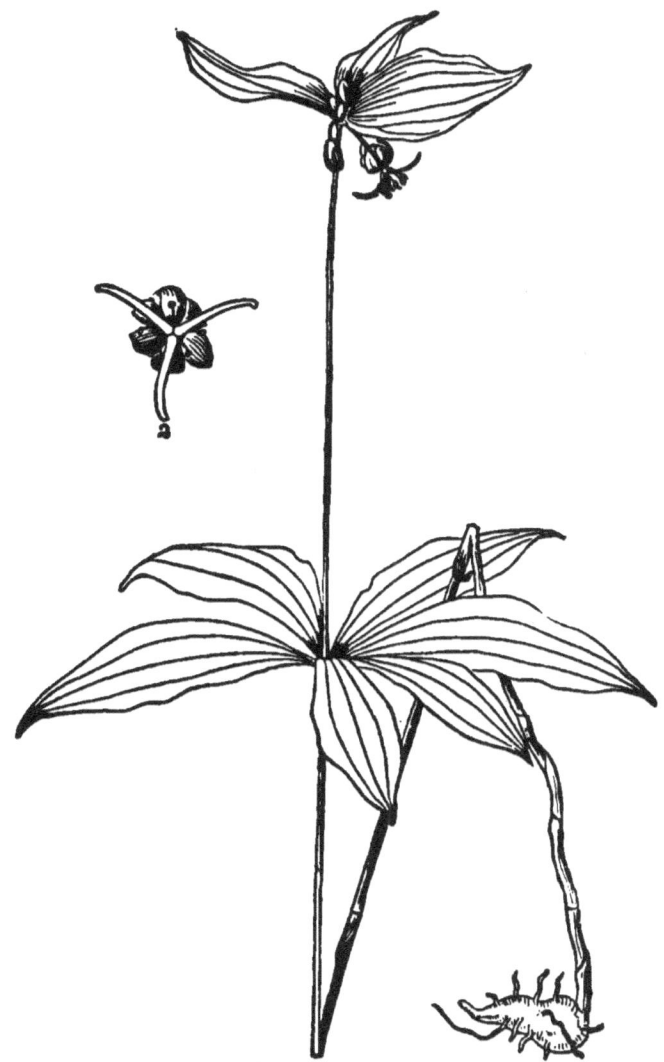

6, Portrait of Medèola Virginica; 2, a flower, life size.

together with 2 or 3 greenish flowers on recurved peduncles. The remarkable feature of the flower is, the very long, reddish stigmas. The student will systematically analyze and record.

The Name is *Medèola Virgínica*—Medeola derived from the fabled sorceress Medea; Virginica, because the plant was first found in Virginia (by Gronovius).

The Order TRILLIACEÆ, represented by these plants, comprehends only 4 genera, and about 30 species. Some authors unite this order to the Lilyworts. Their rhizomes are generally emetic, some of the Trilliums violently so.

LXVI. BELLWORT, OR WILD OATS.

Description.—Associated in memory with babbling brooks, mossy banks, grassy knolls, in the borders of meadow and forest, are the hanging Bellworts, known to our childhood as Wild Oats. They come not in Flora's advance-guard with Bloodroot and Erythronium, but follow later, in May, when her ranks are already full. It would be desirable to study this plant in connection with Erythronium; but as their flowers are not contemporary, a dried specimen, or the analysis (p. 34) must suffice.

Analysis (generic).—Five or six kinds of Bellwort may be found, all flowering in May. Our specimens may therefore be various, yet all smooth and delicate herbs 6–18' high. The stem rises from a rhizome, forks into two branches above, both leafy, and one bearing a drooping flower on a peduncle, which is at first terminal, but becomes axillary by the further development of the branch. The leaves are parallel-veined, oval or oblong, and either *sessile*, or clasping the stem at the base (*amplexicaul*), or *perfoliate*, i. e., with the stem passing through the blade near the base.

The *Perianth* is between cylindric and bell-shaped, consisting of 3 sepals and 3 petals all similar in color and *lance-spatulate* in form, often twisted, having a honey groove or

Fig. LXVI.—Uvulària sessilifòlia: 2, section of the flower; 3. the pistil—triple, 3-parted above; 4, the capsule: 5, a cross-section, showing the 3 cells; 6, section of a seed, with embryo; 7, plan of the flower,—all in 3s and alternating.

pit at base, and deciduous. The anthers of the 6 stamens are adnate, extrorse, and longer than the filaments.

The *Ovary*, as in Erythronium, is triple, and the short style bears 3 long, distinct stigmas. The fruit is also a capsule, but with fewer seeds, and the valves open directly into the cells, that is, they are *loculicidal* (*locula*, a cell, *cido*, I cut). A section of the seed largely magnified shows an embryo with one cotyledon in much albumen.

The Name, *Uvulària*, was conferred on this genus by Linnæus for the fancied resemblance of the pendant flowers to the human palate (*uvula*). The common species, portrayed in Fig. LXVI, is *U. sessilifòlia* (the sessile-leaved), having the leaves sessile, glaucous beneath. The flower is of a creamy white, hardly 1' long, with the styles nearly as long and half united.

U. perfoliàta, the perfoliate-leaved, is also common. The cream-colored flower is more than 1' long, and the petals are covered and roughened inside with grains, or a mealy dust.*

U. grandiflòra, the great-flowered, has also perfoliate leaves, a flower 1¼' long, not mealy inside.

6, Uvulària perfoliàta.

---

* In this species and the next, the nature of perfoliate leaves is seen. The stem passes *through the blade* (*per*, through, *folium*, leaf) near the base. But here the upper leaves gradually become heart-shaped, and the terminal one is nearly sessile, as in *U. sessilifolia*. This shows that these leaves become perfoliate by first growing sessile, then enlarging backwards into base lobes, which finally unite by their inner edges and close around the stem, much as the *peltate* leaves of Tropæolum (p. 91) or the upper (double) leaves of the Honeysuckle,

THE BELLWORT.

ORGAN.	*L*ife, *H*abit, *N*umber, *P*lace, *D*ehiscence, *K*ind, *C*onstruction, *F*orm, *P*lacentation, *S*ize, *Q*ualities, *A*ppendages.
Plant, L.H.S.Q.	♃, *erect, forking above, 10–14', smooth.*
Root, L.K.	♃, *fibers from the joints of a root-stock.*
Stem, L.H.K.F.	*Herbaceous, both branches leafy, one floriferous.*
Leaves, L.P.C.F.S.Q.	*Cauline, alternate, perfoliate, elliptical, 2–3', thin.*
Inflorescence, P.K.A.	*Axillary, solitary, pedunculate, pendulous.*
Flower, N.C.	*One, perfect, complete, 3-parted, hypogynous.*
Calyx, F.Q.	*Like the corolla, pale-yellow, with honey grooves.*
Sepals, L.N.P.F.	*Deciduous, 3, little spreading, linear-oblong.*
Corolla, F.Q.	*Forming with the calyx a bell-shaped perianth.*
Petals, L.N.P.F.	*Deciduous, 3, linear-oblong, granulated within.*
Stamens, N.P.C.	*6, hypogynous, with filaments.*
Anther, D.C.F.	*Longitudinally 2-celled, linear.*
Style, N.C.F.	*One, deeply 3-cleft.*
Stigma, N.F.	*The 3 branches stigmatic along their inner surface.*
Ovary, C.F.Pn.	*3-celled, elliptic-oblong, pn. central.*
Fruit, N.D.K.F.Q.	*3-valved, loculicidal, capsule.*
Seed, N.C.F.Q.A.	*Few, anatropous, obovoid; raphe fungous.*

LOCALITY.—Woods, Akron, O. (Date), May, 1868.

CLASSIFICATION.—FLOWERING ENDOGENS.

—Order, LILIACEÆ. THE LILYWORTS.

NAME.—Latin, **Uvularia perfoliata.**

—English, *Perfoliate Bellwort.**

---

* Our five kinds of Uvularia may be distinguished as follows. (See *Botanist and Florist*, p. 347, Flora.)
Leaves sessile.
  1. *U. sessilifolia.* Leaves glabrous, glaucous beneath. Style ½ 3-parted. Pod 3-angled.
  2. *U. pubérula.* Leaves puberulent, green both sides. Style 3-parted. Pod ovoid.
Leaves perfoliate.
  3. *U. flava.* Perianth 1' long, bright yellow, smooth both sides. Leaves obtuse.
  4. *U. perfoliata.* Perianth 15" long, pale-yellow, covered inside with shining grains.
  5. *U. grandiflora.* Perianth 18" long, smooth, straw-yellow. Anthers obtuse.

LXVII. LILY OF THE VALLEY.

Description.—In May seek also, in the gardens, the Lily of the Valley, exquisite in delicacy and sweetness, and analyze in connection with Uvularia. It is originally a

Fig. LXVII.—Convallària majàlis : 2, section of a flower ; 3, the ripe berry.

mountain plant of Europe and grows wild on the high Alleghanies of Virginia and Carolina. It is propagated by its rhizomes. In the cultivated state it bears no fruit, or but little, perhaps for want of the special insect by which its flowers are fertilized in its native mountains. There its red, round, few-seeded berries are perfected in abundance.

Analysis.—How much of the plant is subterranean? On this large proportion depends its almost unconquerable vitality. The Lily of the Valley is strictly *acaulescent*. From each bud of the running, slender rhizome arise 2 leaves and several bracts involved together,* and a scape outside of them (herein different from Erythronium), bearing a *secund* or one-sided raceme.

The *Perianth* is remarkably distinguished, being strictly *gamopetalous*. Its 6 united leaves are indicated only by the six teeth of the border.—But we are saying more than behooves us. Let the student make thorough inquiry and record of every organ, marking especially the contrasts with Uvularia or Erythronium.

The Name of this plant is *Convallària majàlis;* the generic term being derived from the Latin word for valley, the usual place of growth of some of the species.†

Clintonia. In the coldest woods of the Northern States grows the Yellow Clintonia (Fig. LXVII, 4), flowering in June. The dignity and elegance of its port compensate for its dull colors, and the collector is proud of its discovery.

---

* In the portrait (Fig. LXVII) we seem to have a stem and a peduncle (a, s). But the stem is only the petioles bound together by sheathing bracts. Let these be fused as well as bound together, and they will become a stem indeed. Thus the origin and nature of the stem are clearly indicated. It is formed of the united bases of all the leaves—even the columnar trunk, which lifts on high the organs it bears in order to expose them more thoroughly to the quickening influence of the sun and air. We have already seen that the various appendages of the stem—the bracts and scales, the flower with its several organs, and the fruit, are each but modifications of the leaf; and now we learn that the stem itself, even the woody trunk, is indeed a combination of leaves. Hence the conclusion that the LEAF is the one only type of the whole plant.

† Our Lily of the Valley is often supposed to have been the plant alluded to by Christ when he bade his disciples "Consider the lilies of the field" (Matt. vi., 28). Indeed the plant is called by name in Canticles ii., 1. But no Convallaria is found in the Holy Land. The *Krinon* of the New Testament, rendered "Lily of the field," may have been the red Martagon Lily (Lílium Chalcedónicum), or it may have been a general term referring to the splendid scarlet Anemone (*A. coronàrius*) and Ranunculus (*R. Asiáticus*) which overspread the fields of Palestine.

Analysis.—One soon learns to associate this plant with Convallaria, and to analyze it by a series of comparisons. Let the following points be specially investigated:

4, Clintonia borealis; 5, a berry cut across to show the 2 cells.

The Stem, its habit, form and kind.
The Leaves, their clothing, venation and outline.
The Inflorescence and its appendages.
The Flower, its symmetry, cohesions, form of perianth.
The Fruit, its kind, form, color, cells.
The Seeds—number, contents.

The Name is *Clintònia boreàlis*. The genus was dedicated by Rafinesque to Gov. DeWitt Clinton; boreàlis, is the Latin for northern.

Classification.—The Clintonias, Convallarias, Uvularias, Erythroniums, Tulips, are some of the beautiful creations which constitute the order of the Lilyworts— or LILIACEÆ. The true Lilies, of the genus Lilium, will adorn our fields and gardens in midsummer, and add new luster to this splendid order. Lindley estimates its numbers to be—genera, 147; species, 1200, all combining the following traits:

Leaves parallel-veined, simple.
Flowers regular, perfect, almost always 3-parted.
Perianth free from the ovary, its segments colored alike.
Stamens as many as the segments of the perianth.
Styles wholly or partly united.
Fruit a berry or capsule.
Seeds albuminous, one-cotyledoned.

The **Lilyworts** are chiefly herbs, and natives of temperate climates. The Tropical species are generally shrubs or trees. Besides their pre-eminent beauty, many species are variously useful.

The Tulips, Lilies, Day Lilies, Yuccas, Agapanthus, Star of Bethlehem, and Hyacinth are well known garden flowers.

Crown Imperial (*Fritillària imperiàlis*) is a native of Persia. Its crown consists of a tuft of terminal bracts, from the midst of which droop the large red or yellow flowers. The fœtid bulb is said to be poisonous.

Onions, Leeks, Garlics, are the bulbs of various species of *Allium*. Quamash, an important article of food with the Digger Indians of the far West, is the bulb of *Scilla esculénta*, and several species of *Dichelostémma*. Asparagus is the young shoots of *Aspáragus officindlis*.

Aloes is the dried juice of *Aloe spicàta*, a shrub of S. Africa, and of other species of this genus.

Squills, a valuable medicine, is the dried bulb of *Scilla marítima*, of S. Europe.

Dragon's-blood is a resin exuding from *Dracèna Draco*, a large tree of the Canaries. One specimen has a trunk more than 20 feet in diameter; but its height is inconsiderable.

New Zealand Flax is made of the tenacious fibers of *Phórmium tenax*, a plant resembling a Yucca. The still stronger fibers of *Sensivèra* constitute the African or Bowstring Hemp.

LXVIII. THE STAR-GRASS.

Description.—While the open woodlands glow with the purple of the Wild Geranium, and the meadows are touched with the rainbow tints of the Iris, the humble Star-grass, low down in the drier mold at your feet, unfolds its yellow stars and invites a passing glance.

Analysis.—In securing specimens entire, care is needed, for the solid bulb (the *corm*) lies deep and is anchored by strong, fibrous roots.

The *Stem* is represented by this corm alone.

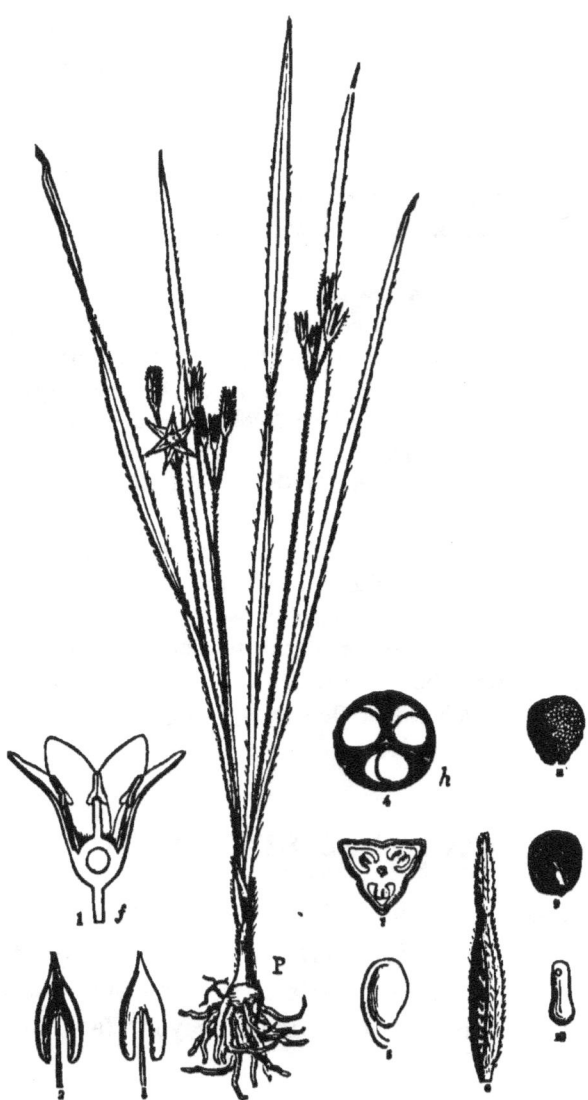

Fig. LXVIII.—Hypóxis erecta: 1, section of a flower; 2, anther seen from within (introrse); 3, anther, outer side; 4, section of the ovary; 5, an ovule inverted on its stalk (anatropous); 6, a capsule partly open; 7, cross-section of the same; 8, a seed; 9, seed dissected; 10, the embryo removed from its albumen.

The *Leaves*, compared with Iris, are both like and unlike. In venation, parallel-veined; in vernation, equitant and *triquetrous* (3-rowed); in outline, linear; in position, vertical. Several of the outer leaves are reduced to mere sheaths involving all the others at the base. In surface character, i. e. in *quality*, the leaf, and indeed the whole plant, is clothed with soft, scattered hairs.

Inflorescence. —Several scapes shorter than the leaves (2–6′, leaves 3–8′) issue with them, and stand erect, although as slender as a thread (filiform), each bearing an irregular umbel of 2–5 flowers. The minute bracts, forming an involucre, must not escape notice.

The *Flower* is perfect, regular, having the usual 4 sets of organs, and closely analogous to both the Irids and the Lilyworts. But from these two orders the Star-grass differs severally by at least one important character. Here let the student close the book, and determine these differences for himself. The sepals and petals being similar form a *perianth*.

The Irids have 3 stamens with extrorse anthers. How is this in the Star-grass?

The Lilyworts have the perianth free from the ovary. How is this in the Star-grass?

Compared with the Trilliads or the Orchids, the differences become many and more obvious. What are they?

Are the sepals and petals quite similar in form and color? Both are imbricated and persistent, withering on the adherent ripening ovary. The 6 anthers are *sagittate*, i. e., arrow-shaped. Only one style appears and one capitate stigma; but the fruit is a 3-celled capsule, containing many roundish, black seeds.

The Name is *Hypóxis erécta*—Hypoxis from two Greek words signifying "sharp beneath," probably referring to the form of the ovary, or flower-bud.

H. filifòlia, the thread-leaved Hypoxis, is another species, prevalent in the Southern States, with filiform leaves as well as scapes, only half a line wide.

LXIX. NARCISSUS.

Description.—In Spain, the Jonquils, Daffodils, Polyanths and Narcissi are wild native plants. In America, they flourish only in gardens under the florist's care, prized for their elegance and sweetness. They begin to bloom a week or two earlier than Hypoxis, with which plant the student will do well to compare them.

Analysis.—Narcissus agrees with Hypoxis in the form, adhesion and æstivation of the perianth, number of stamens, the style, capsule, and other parts, which will be duly

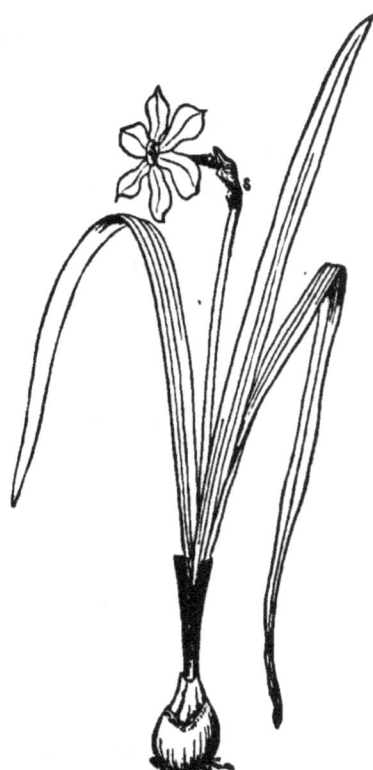

Fig. LXIX.—Narcissus poéticus: *s*, bracts forming the spathe; 2, the flower nearly life size; 3, a flower of N. Pseudo-Narcissus.

recorded. But it differs in its coated bulb (the stem), horizontal, not vertical, leaves, one-leaved, membranous spathe, and most conspicuously in its *coròna*—a cup-shaped appendage crowning the open flower.

The *Corona* arises from the perianth just above the throat, and includes the 6 *unequal* stamens and the 1 style. In the plant before us—known as the Poet's Narcissus, in which the spathe is one-flowered, the corona is saucer-shaped, much shorter than the white perianth, sulphur-yellow, and edged with vermilion.

The Name.—*Narcissus*, the generic name, comes from the Greek, *narkao*, meaning to become numb; for the supposed effect of its fragrance. *N. poéticus* is the Poet's Narcissus, or the species which Ovid intended in his fable of the youth Narcissus, who pined away with love for his own image reflected in the fountain, and at death was changed into a flower.* Among the numerous species are—

N. Pseudo-Narcissus (False Narcissus), the Daffodil, having the large, yellow flower solitary like *N. poeticus*, but the corona is large, bell-shaped, with a notched margin. It is often double. In this state the petals become numerous, each bearing a fragment of the broken corona; but the 3 outer leaves—the sepals—are free.

N. Jonquilla, the Jonquil; very narrow leaves, 2–5 small yellow flowers on each scape, short corona, and very fragrant. The name is a diminutive of *Juncus*, a Rush.

N. Tazétta, Polyanthus; leaves linear, flowers 5–20 white or yellow, crown yellow.

Classification.—By Narcissus and Hypoxis the order of the Amaryllids—AMARYLLIDACEÆ—is introduced, numbering 68 genera and 400 species, characterized as follows:

> Bulbous herbs with scapes and linear leaves.
> Flowers showy, perfect, not woolly nor scurfy.

---

* The ancients used the Poet's Narcissus as a funeral flower, and it was consecrated to the Furies who are fabled to stupefy their victims before punishing them.

Perianth 6-parted, imbricated, adherent.
Stamens 6, anthers introrse.
Ovary 3-celled, with the styles united into one.
Fruit a berry or a capsule.
Seeds one-cotyledoned, albuminous.

The Amaryllids display their chief glories in S. Africa and Brazil. In other countries, they are thinly dispersed as natives, but well represented in gardens and conservatories.

Here belong the fair Snowdrop (*Galánthus*), the graceful Snowflake (*Leucójum*), the splendid Jacobæa (*Sprekélia*), *Amaryllis, Pancrátium, Crìnum*, etc.

The Tuberose (*Poliánthes tuberòsa*, i. e., tuberous-rooted), so powerfully aromatic, is a native of Ceylon.

Hæmánthus toxicárius and other species have poisonous bulbs, used by the Hottentots for poisoning their arrows. The flowers of the Daffodil are said to be poisonous.

The American Aloe or Century Plant (*Agàve Americàna*) is a native of Mexico, well known in cultivation. It is a gigantic herb, flowering but once, after a growth of 50 to 100 years. But then its blossoms are numbered by thousands, panicled on a scape 30 feet in height.* The juice of its immense leaves is mildly acid. By fermentation it is perverted into a vinous beverage resembling cider, except in its nauseous smell, and is much used by the Mexicans under the name of "pulque." The fibers are manufactured into thread. The juice when dried or vaporized becomes a useful soap. A variety in cultivation has its leaves beautifully striped.

LXX. THE SEDGES. GALINGALE.

Description.—The Sedges bear a general resemblance to the Grasses, and are often, by the unlearned, mistaken for them. Both grow in similar situations, but in low, wet lands the Sedges usually prevail. They are generally known by their *solid* (not hollow) stems and *entire* (not split)

---

* Herbs fruiting after a long term of years only once and then dying, are said to be *monocarpic perennials*. Such also is the Talipot Palm of Ceylon.

THE SEDGES.

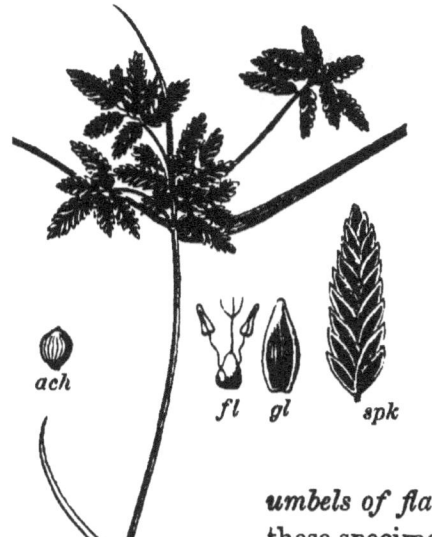

Fig. LXX. — Cypèrus diándrus: *fl*, a flower; *gl*, glume; *Ach.*, achenium.

sheaths.* Being, like the Grasses, almost ubiquitous, specimens may be culled in great variety, during Spring and Summer, in the meadows, fields, open woods, or even in the gardens.

Analysis. — GENERIC CHARACTERS.—Let us begin with the Galingales—a genus of Sedges known at sight by their *terminal umbels of flattened (2-edged) spikes*. In these specimens, fresh or dried, the *culms* (so the peculiar, jointed stems of grass-like plants are called) are triangular, erect, leafy below, solid with pith. The Leaves are linear, parallel-veined, supported on sheaths which are closed around the culm below, never split as in the Grasses.

Inflorescence. The umbel is subtended by an involucre composed of several unequal leaves, and its very unequal *rays* (peduncles) are each *sheathed* at the base.

The *Flowers* occupy the spikes, which

5, a flower of a Rush (Lùzula).

---

* The student will not mistake for Sedges those Rushes which have regular, 3-parted, green flowers with 3 sepals, 3 petals, 6 stamens, 3 stigmas, and several seeds in the capsule, as seen much magnified in the cut.

are composed of imbricated bractlets called *scales* or *glumes* arranged alternately and in 2 rows. Each glume, except the lowest, conceals in its axil one minute, naked flower consisting of a 1-ovuled ovary with 3 (rarely 2) stigmas and 3 (rarely fewer) stamens.

The *Fruit* is an achenium 2 or 3-angled, its seed with a minute embryo in the end of the mealy albumen.

SPECIFIC CHARACTERS.—Fig. LXX portrays a common Sedge, called Brown Galingale. The root is annual and fibrous. The culms rise 4–10′, sheathed and leafy below, naked above, bearing a simple umbel with several rays and an involucre of 3 very unequal leaves. The spikes are clustered on the rays, oblong, obtuse, flat, about 8-flowered, and usually brown in color. The tiny flower beneath each glume except the lowest one has only 2 stamens and 2 stigmas, and finally an achenium 2-edged.

The Name.—This pretty plant represents the genus *Cypèrus*—a genus of immense extent, dedicated to the *Cyprian* queen, Venus, the fabled goddess of beauty.* The species is *C. diándrus*, or the Two-stamened Cyperus, so named by the late Dr. Torrey,† on account of the rare specific character thus denoted.

Scientific Terms.—Culm. Glume. Rays. Scale. Spike.

LXXI. THE SEDGES. CAREX.

Description.—There is scarcely any kind of soil or locality where a Carex may not grow. Look for them in forest, field or meadow, on mountains or prairies, in lands

---

* Cyperus includes not less than 850 species, of which about 40 are natives of the United States.

† John Torrey, M.D., professor of Chemistry at West Point and Princeton, and of Botany in Columbia College, died 1873, æt.77; among American botanists pre-eminent, beloved and revered as an instructor, friend and Christian.

THE SEDGES.

Fig. LXXI.—Carex bullàta: 2, a sterile flower; 3, a fertile flower dissected, showing the glume, ovary and stigmas; 4, section showing the solid culm and equitant vernation.

shady or arid, loamy or rocky, in sands or peat-bogs—anywhere, and the search need not be long. They are readily known from other Sedges by having their flowers all imperfect, either ♀ or ♂, and the achenium inclosed in a bottle-shaped sack.

Analysis.— 1. GENERIC CHARACTERS. — Having in hand a Carex, or any number of them, fresh or dried, their features may be traced as follows: A triangular culm, or a cluster of culms, beset with grass-like leaves, bears one or several, often many, greenish spikes. The spikes are *terete,* composed of *glumes* (or scales) *spirally* imbricated, and bearing in the axil of each glume (except the lower) a single flower.

The *Flowers* are all imperfect, either staminate (♂) or pistillate (♀), and variously disposed. In some species, the ♂ and ♀ together occupy the same spike or spikes (*androgynous*); in other species they occupy separate spikes on the same plant (*monœcious*); and in a few others, separate spikes on separate plants (*diœcious*). The ♂ flowers consist of 3 stamens, with anthers attached to the filament by the base, i. e., *innate*.

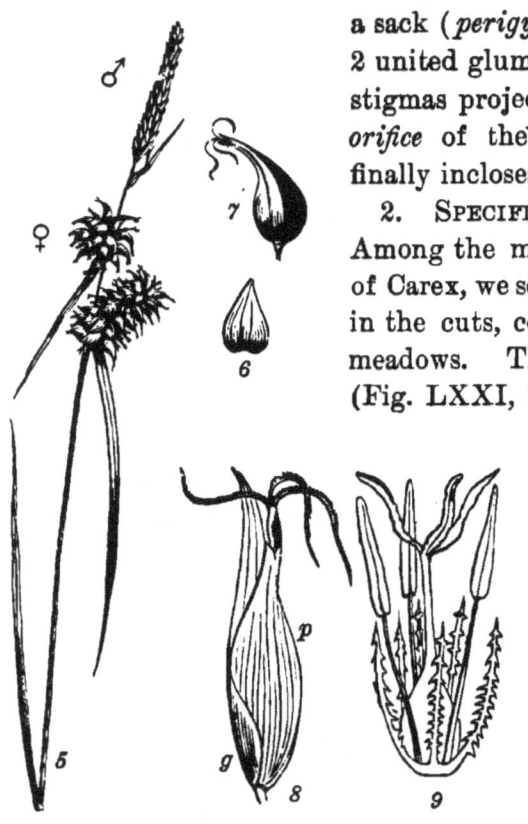

The ♀ is an ovary invested with a sack (*perigynium*) composed of 2 united glumes. The 2 or three stigmas project from the beak or *orifice* of the perigynium which finally incloses the achenium.

2. SPECIFIC CHARACTERS.—Among the multitudinous forms of Carex, we select the two shown in the cuts, common in our wet meadows. The Jewelled Carex (Fig. LXXI, 1), may be distinguished thus: A smooth, light-green Carex, 2 feet high, with narrow leaves and bracts, monœcious, with the sterile (♂) spikes 2 or 3, and the fertile (♀) 1 or 2, oval or oblong, on very short peduncles; the perigynia turgid-ovoid, reclining, tapering into a long, straight, rough beak, much longer than the lanceolate glume; stigmas 3, achenium 3-cornered.

5, Carex flava; 6, a glume; 7, a flower (perigynium) with 3 stigmas issuing from the orifice; 8, ♀ flower of Carex rivulàris; *g*, the glume, *p*, the bottle-shaped perigynium 2-toothed at top, enveloping the ovary; stigmas 3; 9, a perfect flower of Scirpus lacustris, with 6 setæ, 3 stamens, 3 stigmas.

The Yellow Carex is thus distinguished:

Inflorescence monœcious, ♂ spike single, ♀ spikes 2.
Stigmas 3, and the achenium therefore 3-cornered.
Peduncles (♀) scarcely exserted from the sheaths.
Perigynia smooth, crowded, inflated, longer than the glume, ovoid, tapering into a slender *recurved* beak.
The whole plant is yellowish-green, 10–20' high. The ♂ spike is terminal, cylindric, 10–12" long, the ♀ spikes roundish to oval, 4–7" long.

The Name, *Carex,* is the old Latin name for these plants,* from *careo,* I want; as the upper flowers are constantly without seed. *C. bullàta,* the Jewelled Carex, is named for its stud-like perigynia; *C. flava,* the Yellow Carex, alluding to the yellowish herbage. Carex is the largest genus in the Flora of North America.†

Classification.—These examples must suffice to represent the great order of the Sedges—the CYPERACEÆ, known by the following traits:

> Culms solid with pith.
> Leaves linear, channelled, with closed sheaths.
> Flowers spicate, one in the axil of each glume.
> Perianth none, or a few *setæ,* or a *perigynium.*
> Anthers generally 3, fixed by the base (*innate*).
> Pistil 1-ovuled, with 2 or 3 stigmas.
> Fruit an achenium 2-edged or 3-cornered.

The Order of the Sedges includes 120 genera, and 2000 species. They inhabit all climes and countries, but chiefly the meadows, marshes and swamps of the temperate zones. They are of slight use as food, or in the arts. They differ from the grasses in having little sugar or starch, and so form a poor pasture.

The Nut-grass of the S. States, the pest of the Cotton-fields, is *Cypèrus Hydra.* It multiplies by creeping roots and tubers in spite of hoe and plough. The tubers of *C. esculentus,* cultivated by the ancient Egyptians, may be boiled for food, or roasted for Coffee. The roots of *C.*

---

* See Virg. Ecl., III, 20. "Tu post *carecta* latebras." You hid behind the Sedges. The English term Sedge comes from the Saxon *sæcg,* a sword.

† About 500 species have been described, of which 200 are natives of the United States.

rotúndus contain an aromatic oil; those of *C. longus* are tonic and astringent.

The Mat-grasses, growing on sandy shores and dikes, securing them against the incursions of the sea or the drifting winds, include some species of Carex, as *C. arendria, C. fœnea, C. ripária*, etc. The roots of *C. arendria* are used as a substitute for Sarsaparilla.

ORGAN.	Life, Habit, Number, Place, Dehiscence, Kind, Construction, Form, Size, Qualities, Appendages.
Plant, L.H.S.Q.	♃, *grass-like, 20-30' high, light-green.*
Root, L.Q.	♃, *numerous fibers from creeping rhizomes.*
Culm, L.H.F.S.Q.	*Herbaceous, erect, triang., solid, smoothish.*
Leaves, L.P.C.F.Q.	*Alternate, equitant at base, linear.*
Sheaths, C.S.Q.	*Clasping the culm with joined edges, smooth.*
Inflorescence, P.K.F.A.	*Spikes terminal, and in the upper axils, mostly on peduncles, with bracts, and terete.*
Involucre, N.K.S.	*None.*
Spikes, N.K.F.S.Q.	♂, *2 or 3, terete, acute, 1' and less;* ♀, *1 or 2 below the* ♂, *oval or oblong, short-pedunculate.*
Flowers, N.K.C.	*Many,* ♀ *and* ♂, *the sterile naked (no perianth).*
Glumes, N.P.F.	*One subtending each flower, ovate.*
Perianth, N.P.K.S.Q.	*Perigynium turgid, abruptly long-beaked.*
Anthers, N.C.D.	*3, innate, linear, 2-celled longitudinally.*
Stigmas, N.F.	*3, issuing from the 2-toothed orifice.*
Grain, K.F.A.	*Achenium triangular-obovoid.*
Seed, N.K.C.	*One.*

LOCALITY.—*River banks, Fordham, N. Y.* (Date), *June, 1878.*
CLASSIFICATION.—GLUMACEOUS EXOGENS.
—Order, CYPERACEÆ, THE SEDGES.
NAME.—Latin, **Carex bullata.**
—English, *The Jewelled Carex.*
REMARKS.—*The perigynium is twice longer than the glume.*

THE SEDGES.

Bulrushes, used in making matting, chair-bottoms and baskets, are the culms of *Scirpus lacustris* and other kinds.

The Tule, or Giant Rush, growing in inundated places, is *Scirpus válidus*. In the valley of the Sacramento, Cal., it rises 12 feet high, covering thousands of acres.

The Cotton Grass (*Erióphorum*) is conspicuous in our wet northern meadows for its airy cotton-like tufts waving in the wind. These tufts

6, Cypèrus Papỹrus. A scene in ancient Thebes, on the River Nile.

are composed of the long hairs, called *setœ*, growing in each of the crowded flowers, in the place of a perianth. Five species are described in our floras (*Botanist and Florist*, p. 362), of which *E. Virgínicum*, with reddish cotton, displays the largest tufts.

The Rush or Bulrush of the Nile (Hebrew, *Gome*, Exodus ii., 3) is *Papỹrus antiquòrum* (Willd.), or as now called, *Cypèrus Papỹrus* (Linn.).

It is a gigantic Sedge, 10–15 feet high, surmounted by a compound umbel of numerous rays and bracts. Its spreading rhizomes have helped to consolidate the mud of the Delta. Its tall, stout culms were used in making boats, baskets, ropes, and fuel, as the name implies (Gr. *pao*, to feed, *pyr*, fire). The earliest and rudest paper (hence the name) was manufactured from its pith—the cellular tissue which fills its culms.* Its graceful form affords a favorite theme for artists.

Scientific Terms.—Androgynous. Beak. Equitant. Glumes. Innate anther. Monœcious. Orifice. Perigynium. Spikes.

LXXII. THE GRASSES.

Description.—These modest and useful plants are everywhere at hand, mantling the hills, meadows, and valleys with their soft, uniform green, beginning to open their colorless flowers early in June or sooner. A variety of such specimens is before us to-day, unpromising indeed. But we cannot fail to find the examination full of profit and agreeable surprise. Our cuts represent three common Grasses, with flowers simultaneous, and quite dissimilar.

Analysis (generic).—The *Root* of all these is *inaxial* (no tap root), consisting of many strong fibers taking a firm hold of the soil and helping to bind it into a matted turf.

The *Stem* is somewhat enlarged or bulbous at the base, terete above, conspicuously jointed at intervals, hollow or *fistular* between the joints. Stems of this kind are called *culms*.

The *Leaves* are alternate, parallel-veined, constructed in 3 parts. The lower part, from the joint upward, is the *sheath*, answering to petiole, enfolding the stem with edges overlapping beyond—not united. The blade is strictly linear. At the junction of the sheath and the blade is a short membrane called *ligule*, answering to stipules.

---

* The mode of preparing this paper was very simple. The stem was peeled and the pith cut lengthwise into thin slices. These were then laid side by side with their edges touching, and sprinkled with the muddy water of the Nile. Another row of pith-slices was then laid transversely upon the first, and by pressure the whole mass united into a compact sheet.

The *Inflorescence* is variously developed in these specimens. Generally it appears as in Spear Grass (Fig. LXXII), a branching pyramidal bouquet—a *panicle*, differing from a raceme inasmuch as the branches are branched again. It is well to observe whether the branches are grouped in 5s, 2s, or 1s. In other specimens the inflorescence is contracted into a spike or a spike-like panicle. The flowers are collected into little clusters called *spikelets*. Let us here take up a spikelet of Spear-grass (Fig. LXXII,1), which we may study as a type of all. It is scarcely 2″ in length, ovate-lanceolate

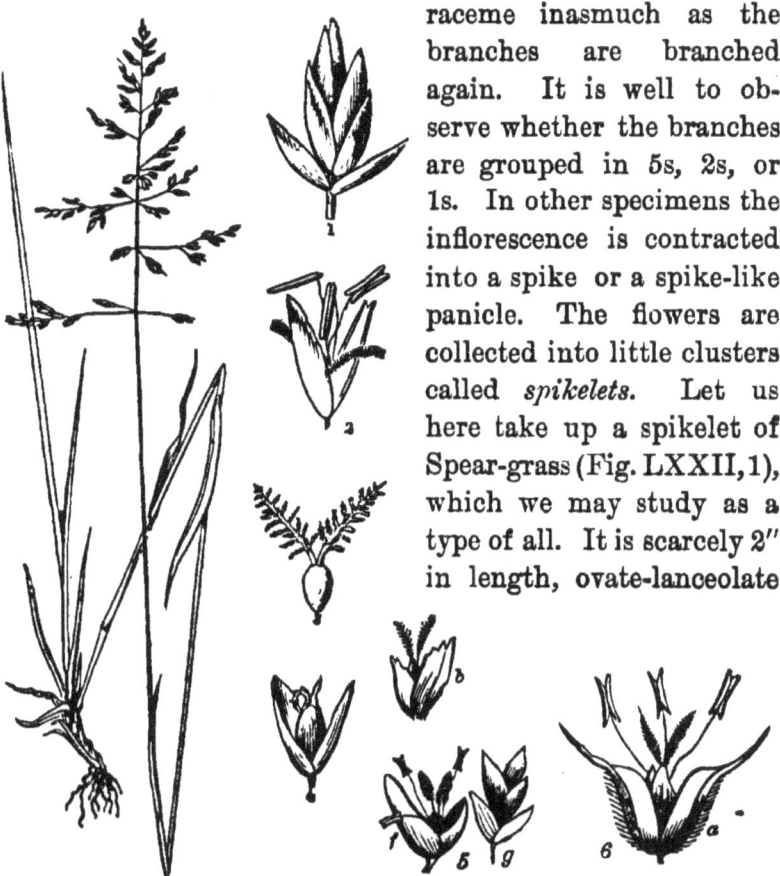

FIG. LXXII.—Poa praténsis (a depauperate specimen, for the branches are usually in 5s): 1, a spikelet with 2 glumes and 4 flowers; 2, a single flower; 3, ovary and feathery stigmas; 4, ripe kernel enclosed in the 2 pales; 5, Poa débilis; *g*, spikelet, 3-flowered; *f*, a flower; 6, a spikelet of Phlèum praténse: *a*, the 2 awned glumes; *b*, the 2 pales and ovary.

in outline. At the base are 2 chaffy bractlets—the *glumes* (*g*). Within and above the glumes are 4 flowers (more or less), alternate, imbricated when closed. Such is a spikelet,

ORGAN.	Life, Habit, Number, Place, Kind, Construction, Form, Size, Dehiscence, Qualities, Appendages.
Plant, L.H.S.Q.	♃, *grass-like, 2-3 feet, dark green.*
Root, L.K.	♃, *many branching fibers.*
Culm, L.H.F.S.Q.	*Herbaceous, erect, simple, terete, hollow, smooth.*
Leaves, L.P.C.F.Q.	*Alternate, flat, broad-linear, rough-edged.*
Sheath, C.S.Q.	*Loose, with edges free, smooth, striate.*
Ligule, F.S.Q.	*Very short or obsolete.*
Inflorescence, P.K.F.A.	*Panicle somewhat secund, slender, 9', the branches short.*
Spikelet, N.K.F.S.Q.	*Fewer, lanceolate, terete, acute, 7″-8″.*
Flowers, N.K.	*6-9, perfect.*
Glumes, N.P.F.S.Q.	*2, upper one larger, 3-veined, scarious on the margin.*
Pales, N.P.K.Q.	*2, the lower one 5-veined, acute.*
Awns, P.F.S.Q.	*None.*
Rudiments, K.F.Q.	*None.*
Anthers, N.P.C.D.	*3, exserted, 2-celled, versatile, longitudinal.*
Stigmas, N.C.F.	*2, plumous, shorter than pales.*
Grain, K.F.	*Caryopsis, oblong.*
Seed, N.K.C.	*One.*

LOCALITY.—*Meadows, Chester, Penn.* (Date), *June, 1876.*

CLASSIFICATION.—GLUMACEOUS ENDOGENS.

—Order, GRAMINEÆ, THE GRASSES.

NAME.—Latin, *Festuca pratensis.*

—English, *Meadow Fescue.*

REMARKS.— *Spikelets somewhat racemed in the branches.*

The *Flower*. A common lens will be helpful in viewing the blossom. Take it while it is open, or in bloom, as many are every dewy morning. First, 2 chaffy bractlets, the *pales,* are seen expanded, one a little above the other, as in the glumes; next, 3 stamens with gossamer filaments and *versatile* anthers; lastly, an ovary with 2 feathery stigmas. In a few days the ovary is matured into a *caryopsis*—a one-seeded fruit like a grain of wheat, whose shell or pericarp is inseparable from the seed.

Fertilization.—There are no bright colors in these flowers to catch the eye of the insect tribes, nor honey to attract them. Insect aid in fertilization does not here seem necessary. The pollen is conveyed by the wind. To this end, the Grasses grow together in dense crowds and the pollen is superabundant, probably a thousandfold, filling the breeze so that the plume of every stigma is sure to catch at least one grain either from its own or other anthers.

The Name, *Poa praténsis*—Poa, Gr. for hay or fodder; praténsis, of the meadow. This plant is generally known as June Grass, as its grains are often ripe in June. We have other species of Poa, flowering a month later, among which is the Blue Grass (*P. compréssa*), also the beautiful red-tinted Fowl-Meadow (*P. serótina*). *P. ánnua* is the low, soft Lawn-grass, flowering in April.*

Scientific Terms.—Caryopsis. Culm. Glumes. Inaxial root. Ligule. Pales. Panicle. Sheath. Spikelets.

LXXIII. THE ORCHARD GRASS.

Description.—This is a conspicuous and very common herb in orchards and groves. It is tall and stout compared with Spear-grass; in color, glaucus or seagreen.

---

* The Annual Meadow Grass (*P. ánnua*) and Shepherd's Purse are, perhaps, the most common plants in the world. On almost every waste spot where even a weed can grow—on the bank by the roadside, along the garden path, between the stones of the city pavement, high up in the mountain as well as in the rich meadow at its foot —these modest plants display their cheerful verdure.

Analysis.—The student will analyze throughout as we have done in Spear-grass, making special note of the points of difference in the two plants, such as the following.

The Orchard-grass is rough to the touch—*scabrous*. The leaves and even the sheaths are decidedly keeled (*carinate*, or boat-shaped).

The ligule is excessively large, and split or bifid.

The branches of the panicle are single.

The spikelets are collected in dense, one-sided (*secund*) clusters. While closed they are lanceolate, 2–3″ long, about 4-flowered, with the flowers a little separated on the rachis. The 2 glumes and the lower pale are rough-*ciliate* (with a row of short hairs) on the keels, and narrowed to an awn-like point. There are 3 stamens, 2 feathery stigmas, and a lanceolate grain free from the pales.

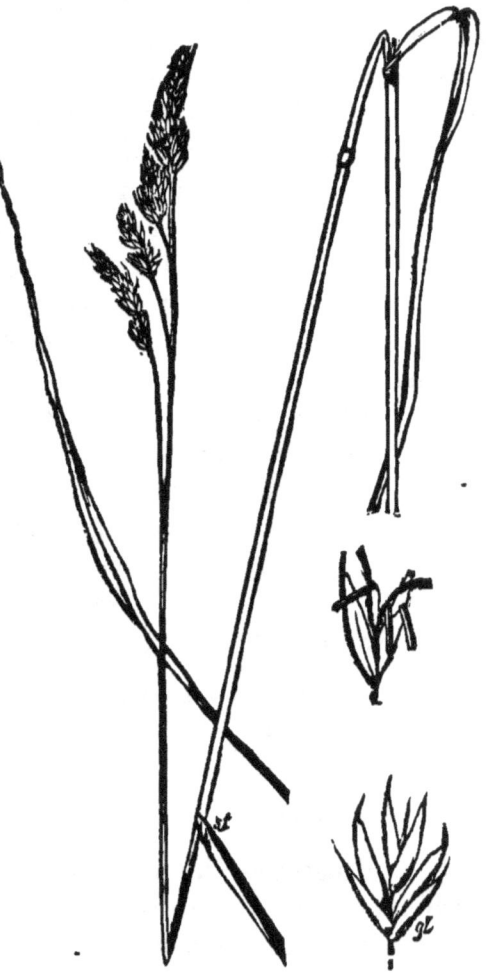

Fig. LXXIII.—Dáctylis glomeràta: 1, a spikelet with 2 glumes, 4 flowers and 1 rudiment; 2, a flower.

The Name.—*Dáctylis glomeràta* (*Dactylis*, fingers, *glomerata*, crowded), is the significant title.

Scientific Terms.—Carinate. Ciliate. Secund. Scabrous.

LXXIV. SWEET VERNAL GRASS.

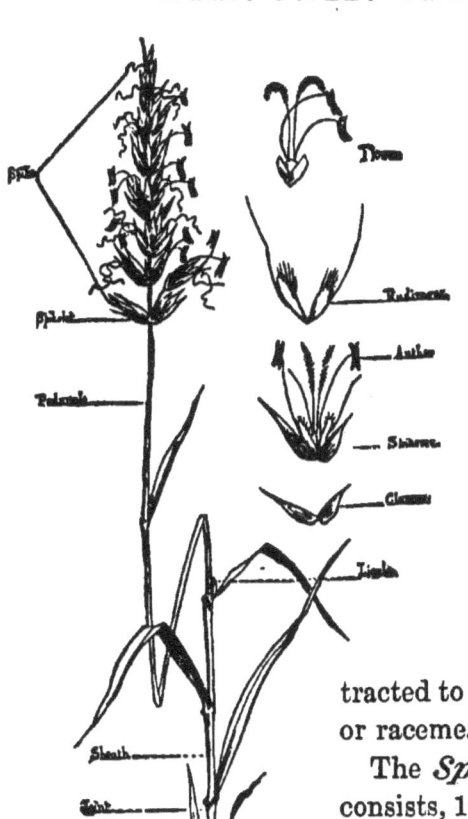

Fig. LXXIV.— *Anthoxánthum odorátum.*

The Sweet Vernal Grass begins to flower a few days earlier than Spear-grass, in the same situations. The analysis of its root, culm, leaves and inflorescence may be conducted as in that plant, searching for differences as well as analogies.

Analysis.—The leaf-blade and sheath are very short, while the internode is very long. The ligule is conspicuous. The panicle is contracted to the form of a loose spike or raceme.

The *Spikelet* is 3–4″ long. It consists, 1st, of 2 glumes, the lower twice longer than the upper; 2d, of ciliated pales supposed to be 2 rudimentary flowers, each bearing an awn on its back; 3d, one perfect flower situated between the 2 rudiments, that is, terminal.

The *Flower* is *diandrous*, composed of 2 small, smooth pales, 2 long stamens with versatile double anthers, and an ovary with 2 styles bearing plumous stigmas as long as the stamens.* Finally the grain or caryopsis resembles a wheat kernel, but many times smaller.

The Name.—To this fine Grass, which is widely dispersed over Europe as well as America, Linnæus gave the name of *Anthoxánthum odorátum*, meaning either "Sweet-smelling yellow flowers" or "Sweet-smelling flower of flowers." But its flowers are not sweet-smelling, yet its herbage when cut imparts to the drying hay much of its delicious fragrance.

Thus we have analyzed three genera of Grasses. In addition, let the student study the Red-top, whose spikelets are simply 1-flowered; Wheat, Oats, and Corn. In the latter the flowers of the tassel are all staminate; of the ear all pistillate.

2, Agróstis vulgàris, *a*, 1-flowered spikelet; *b*, the flower removed from its glumes; 3, Agróstis scabra; *c*, the 2 glumes separated from (*d*) the single flower.

Germination. In the Exogens, as we have often seen, the embryo of the seed has two lobes or cotyledons, or as botanists say, is dicotyledonous. In the Endogens, the embryo is more simple, being generally an oblong body (Fig. LXXVIII, 10), of which one end is a radicle and the other a plumule wrapped up in a single cotyledon, only its end being visible. The nourishment is partly in the cotyledon and mostly the mealy albumen on one side of it. In germina-

---

* We cannot fail to observe the special adaptation of these flowers to wind-fertilization. Their long exserted stamens and stigmas are lifted to the breeze like waving banners. The anthers opening their whole length, swing nicely balanced and tremulous, while the stigmas wave their plumes to catch the flying pollen grains.

tion the cotyledon never arises above the ground but remains with the seed.

Let a few kernels of corn be placed on a lock of cotton in a glass of water. After a day or two the albumen has softened, swelled, and become sweetish. In 3 days the radicle (*r*) has pushed out and turned downward, while the cotyledon has extended itself backward a little, and freed the end of the plumule, but still holding fast above to the albumen whence yet comes its nourishment. Another day, the plumule pushes out from the cotyledon, and begins to ascend as the first leaf, while the radicle grows and develops some side rootlets. Another day we see a second leaf push out from the first, both still rolled up (convolute), and so on, one after another, in the order of a spiral.

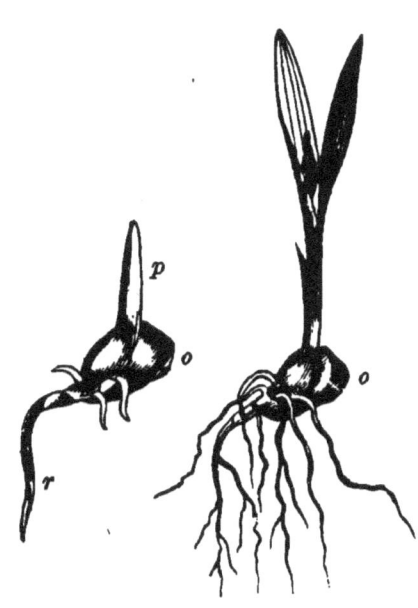

. 4, Germination of Maize (or Indian Corn): *o*, the kernel or albumen ; *r*, the radicle ; *p*, plumule.

Classification.—The order of the Grasses—GRAMINEÆ—includes 300 genera and nearly 4000 species, and limited as follows :

> Plants endogenous, with fibrous roots.
> Culms hollow between the solid joints.
> Leaves alternate, on open or split sheaths, with a ligule.
> Flowers in spikelets with alternate glumes and pales.
> Stamens 3 (rarely 1-6), exserted, versatile.
> Ovary with 2 plumous stigmas and one ovule.
> Fruit a caryopsis, with flowery albumen.

The Order of the Grasses prevails wherever there is a soil, in all countries and climes, varying in species and habit with the climate. In cool, temperate regions, they are dwarfed and crowded, forming a soft carpet of turf. In

6, creeping rhizome of Witch Grass (*Triticum repens*).

warm or torrid regions they form no turf, but grow apart, enlarged, even arising into trees like the stately Bamboo.

The order is no less valuable to man than vast in extent. It furnishes sweet herbage as fodder for animals, and farinaceous grains as food for man. No poisonous herb is found among them except perhaps the Darnel.*

The Common Oat, *Avèna sativa*, is said to be a native of the Island of Juan Fernandez. Its grain is richly nutritious both for man and beast. The grain is firmly inclosed in its husk—the pales—from which it is first separated in the manufacture of *oat-meal*. The Wild Oat, or Animated Oat, is *Avèna fátua*. Its long hygrometric awn is put in motion by slight changes in the moisture of the air.

Barley is the grain of *Hórdeum distichum* and *H. vulgàre*, the former the Two-rowed, the latter the Four-rowed. Native country unknown.

Rye (*Secàle cereàle*) is next to Wheat in value for bread-making. It is chiefly cultivated in Russia and Germany.

---

* Said to be a narcotic poison; but this has not been fully proved. It is the same plant (*Lolium temulentum*) alluded to in St. Matthew as *Tares*. Some Grasses with creeping subterranean stems, as Quick Grass (*Triticum repens*) are troublesome weeds. Others of similar habit, as *Elymus arenarius*, *Psamma arenaria*, bind the loose sand of the sea-shore together, preventing incursions of the sea. Numerous floating islands in the River Amazon are formed chiefly of Grasses. They are called *Llhas de Capim*. Some of these islands are acres in extent, and from 5 to 8 feet of their thickness is under water. (See *Harper's Magazine*, February, 1879.)

Wheat (*Triticum sativum*), as the food of civilized man, is the most valuable of all grains. It has been so long and so extensively cultivated that it has sported into innumerable varieties, and its nativity is unknown.

Rice (*Oryza sativa*) is said to be the food of a majority of the human race, being the staple diet of China and the East Indies. It is largely cultivated in S. Carolina, Georgia and Florida, in lands inundated for this purpose.

Indian Rice, or Wild Rice, is the grain of *Zizània aquática* of river marshes in Canada and the northern United States.

Indian Corn (*Zea mays*) is a native of America, unknown to Europeans until the discovery of the New World. The vast extent to which it is now cultivated indicates its value. Among its many varieties are Sweet Corn, Pop Corn, and Rice Corn.

Sugar Cane (*Sacchàrum officinàrum*) is an important member of this Order, cultivated in warm climates. It is the source of nearly all the sugar consumed by civilized nations. The juice is expressed from the stalks and evaporated until crystallized.

Various species of Sorghum have been cultivated for sugar with poor success.

Broom Corn is *Sorghum sacchardtum*, a native of Arabia. Its use in broom-making is well understood.

Pampas Grass (*Gynèrium argênteum*) is native of S. America, cultivated for its splendid plume-like panicles of silvery whiteness.

Bamboo (*Bambùsa arundinàcea*), growing in all tropical lands, attains a height of 50 to 80 feet, with a culm 10 inches thick—trees of exceeding beauty and grace. Bamboos are variously useful for "fishing-rods, water-pipes, trellis-work, scaffolding, sails, umbrellas, hats, shields, baskets, ropes, paper."

Hay is the dried herbage of many grasses cut when in or just past flowering. Some of the best for this purpose are the following : Timothy (*Phleum praténse*); Redtop (*Agróstis vulgàris*); Bluejoint (*Calamagróstis Canadénsis*); Orchard Grass (*Dáctylis glomeràta*); Spear or June Grass (*Poa praténsis*); Fowl Meadow (*Poa seròtina*); Fescue (*Festùca praténsis*) ; Blue Grass (*Poa compréssa*), Chess (*Bròmus secàle*) ; Salt Meadow Grass (*Bryzopyrum spicàtum*).

Scientific Terms.—Awns. Rudimentary. Versatile anthers.

Appendix.

APPENDIX.

CLASSIFICATION.

In the foregoing lessons we have hastily traversed the Vegetable World, pausing here and there, in such tribes and families as offered special attractions, to gain information more definite and minute. We have observed that while there is found in every species some one fact or principle peculiar to itself alone, yet each plant bears a resemblance, greater or less, to every other plant, so that a bond of affinity pervades the entire kingdom, combining all into one grand system, which to interpret aright is the glory of Man.

Species and Genera.—The individual plants constituting the Vegetable World, so vast and incomprehensible in multitude, are, as we have seen, assorted by nature into species. A *species* may be defined as *a group endowed with the power of perpetuating its own kind and no other*, and thus is maintained the same from age to age. Again, the species themselves, by their mutual resemblances, are grouped into genera. A *genus* is defined as a group of closely related species, *having more resemblances than differences.*

Orders.—The third step in Classification is the formation of orders. As species are grouped into genera, so the genera are collected into orders. An *order* may be defined, *an assemblage of related genera.* The orders differ greatly in respect to their extent, some including few genera, or even but a single genus (as Platanaceæ, the Plane-trees), while others comprehend a hundred, or even a thousand, as Umbelliferæ, and Compositæ.

Cohorts, etc.—Again, the orders, by some one or few traits which they possess in common, are marshaled into Cohorts; the cohorts into Classes, and the classes into Provinces.

Subkingdoms.—Viewed as a whole, the Vegetable Kingdom subsists in two grand divisions, called the Phenogamia and Cryptogamia, as first shown by John Ray of England, 1682. This division depends on the habitual presence or absence of visible flowers, and is confirmed by important differences of internal structure, and of seeds. (See pp. 16, 27, 43.)

Provinces.—Each subkingdom is again divided into two provinces. 1st. The province Exogens includes all such flowering plants as have seeds with 2 or more cotyledons in the embryo, wood growing by external layers, leaves net-veined, and their flowers never (or very rarely) 3-parted. 2d. The province Endogens includes all flowering plants which have seeds with one cotyledon only, wood growing by internal accretions, leaves parallel-veined, and flowers habitually 3-parted. 3d. The province Acrogens includes all flowerless plants with stems growing in length, or at the point (*akros*) only; as Ferns and Mosses. 4th. The province Thallogens, is composed of plants of the lowest orders, growing indefinitely in shapeless expansions with no distinct axis; as Lichens.

Classes.—The subdivisions of the provinces are called Classes; there are two of each. 1st. The Exogenous Angiosperms have flowers with stigmas, and seed-vessels inclosing the seeds. 2d. The Exogenous Gymnosperms are naked-seeded, having neither stigmas, nor pericarp; as Pinus.

3d. The Endogenous Petaliferæ have petals in their flowers, while (4th) the Endogenous Glumiferæ have green glumes and pales instead of petals and sepals ; as Grasses.

Cohorts are subdivisions of the Classes. Of these the Flowering Plants include seven. A, the Polypetalous Exogens, with corolla of distinct petals. B, the Gamopetalous Exogens, with a corolla of united petals. C, the Apetalous Exogens, without petals and often also without sepals. D, the Conoids (the same as the Gymnosperms), seeds naked, and borne in cones. E, the Spadiciflorræ, Endogens with the flowers on a spadix involved in a spathe (example, Calla). F, the Florideæ, Endogens with flowers in racemes, umbels, etc., not on a spadix. G, the Graminoids or Grass-like plants (same as the Glumiferæ).

The mutual relations of all the above divisions will be seen in the following table, which will also instantly indicate the Cohort to which any plant in hand may belong :

THE VEGETABLE KINGDOM.

The Sub-kingdom, PHENOGAMIA, or FLOWERING PLANTS.

Province 1st.—*Exogens* or *Dicotyledons*. Wood (if any) in layers. Leaves net-veined. Flowers not completely 3-parted.

Class I.—Angiosperms. Flowers with stigmas. Seeds inclosed in seed-vessels.

Cohort A.—Polypetalous Exogens. Corolla with the petals distinct. Example, Roseworts.
Cohort B.—Gamopetalous Exogens. Corolla with the petals united. Example, Bindweeds.
Cohort C.—Apetalous Exogens. Corolla wanting, and often the calyx also. Example, Knotweeds.

Class II.—Gymnosperms. Flowers without calyx, corolla, stigmas, or ovary, and seeds naked.

Cohort D.—Conoids, or cone-bearers. (The same as the Gymnosperms.) Example, Pines.

Province 2d.—*Endogens*, or *Monocotyledons*. Wood growing by internal accretions. Leaves parallel-veined. Flowers [3-parted.

Class III.—Petaliferæ, or Aglumaceæ. Flowers with petals, and no glumes.

Cohort E.—Spadiciflora. Flowers on a spadix, mostly involved in a spathe. Example, Aroids.
Cohort F.—Florideæ. Flowers displayed in racemes, umbels, etc., not on a spadix. Example, Lilyworts.

Class IV.—Glumiferæ. Flowers with glumes and pales—no petals.

Cohort G.—Grainmoids, the Grass-like plants. (The same as the Glumiferæ.)

The Sub-kingdom CRYPTOGAMIA, or the FLOWERLESS PLANTS.

Province 3d.—*Acrogens*, or Point-growers. Stem and leaves distinguishable. Examples, Ferns, Mosses.
Province 4th.—*Thallogens*, or Mass-growers. Stem and leaves undistinguishable. Examples, Lichens, Fungi.

THE NATURAL ORDERS,

The North American chiefly—with familiar examples—their extent in Genera and Species, and numbered as in the FLORA ATLANTICA, *or* BOTANIST AND FLORIST.

COHORT A. THE POLYPETALÆ.

					GENERA.	SPECIES.
1.	RANUNCULACEÆ.	THE CROWFOOTS.	*Ranunculus.*	Bulbous Buttercup.	55	1100
2.	MAGNOLIACEÆ.	MAGNOLIADS.	*Liriodendron..*	Tulip Tree.	9	80
4.	ANONACEÆ.	ANONADS.	*Asimina.*	Papaw.	40	350
5.	MENISPERMACEÆ.	MENISPERMADS.	*Menispermum.*	Moonseed.	22	300
6.	BERBERIDACEÆ.	BERBERIDS.	*Berberis.*	Berberry.	19	140
7.	NYMPHÆACEÆ.	NYMPHIADS.	*Nymphæa.*	Water Lily.	8	55
8.	SARRACENIACEÆ.	WATER PITCHERS.	*Sarracenia.*	Pitcher Plant.	3	8
9.	PAPAVERACEÆ.	POPPYWORTS.	*Sanguinaria.*	Bloodroot.	24	290
11.	CRUCIFEREÆ.	CRUCIFERS.	*Brassica.*	Mustard.	172	1600
12.	CAPPARIDACEÆ.	CAPPARIDS.	*Cleome.*	Spider-flower.	23	340
13.	RESEDACEÆ.	MIGNIONETTE.	*Reseda.*	Mignonette.	6	41
14.	VIOLACEÆ.	VIOLETS.	*Viola.*	Pansy.	21	300
15.	CISTACEÆ.	ROCK ROSES.	*Lechea.*	Pinweed.	7	185
16.	HYPERICACEÆ.	ST. JOHN'S-WORTS.	*Hypericum.*	St. John's wort.	8	250
17.	DROSERACEÆ.	SUN-DEWS.	*Drosera.*	Sun-dew.	6	90
19.	CARYOPHYLLACEÆ.	PINKWORTS.	*Dianthus.*	Pink.	35	1000
20.	PORTULACACEÆ.	PURSLANES.	*Claytonia.*	Spring Beauty.	15	
23.	MALVACEÆ.	MALLOWS.	*Malva.*	Mallow.	59	1100

NATURAL ORDERS.

						GENERA.	SPECIES.
25.	TILIACEÆ.	LINDENBLOOMS.	*Tilia.*	The Linden Tree.		40	350
26.	CAMELLIACEÆ.	TEAWORTS.	*Gordonia.*	Loblolly-Bay.		32	130
28.	LINACEÆ.	FLAXWORTS.	*Linum.*	Flax.		14	140
30.	GERANIACEÆ.	GERANIA.	*Oxalis.*	Wood Sorrel.		16	700
31.	RUTACEÆ.	RUEWORTS.	*Xanthoxylum.*	Prickly Ash.		83	650
32.	AURANTIACEÆ.	ORANGEWORTS.	*Citrus.*	Orange.			
36.	ANACARDIACEÆ.	SUMACS.	*Rhus.*	Sumac.		42	480
37.	SAPINDACEÆ.	MAPLEWORTS.	*Acer.*	Maple.		73	650
38.	CELASTRACEÆ.	STAFF TREES.	*Euonymus.*	Burning Bush.		89	400
40.	RHAMNACEÆ.	BUCKTHORNS.	*Ceanothus.*	Jersey Tea.		37	430
41.	VITACEÆ.	VINES.	*Vitis.*	Grape vines.		3	200
42.	POLYGALACEÆ.	MILKWORTS.	*Polygala.*	Milkwort.		15	500
43.	LEGUMINOSÆ.	LEGUMINOUS PLANTS.	*Robinia.*	Locust.		399	6500
44.	ROSACEÆ.	ROSEWORTS.	*Spirea.*	Hardhack.		71	1000
45.	SAXIFRAGACEÆ.	SAXIFRAGES.	*Ribes.*	Currants.		73	900
46.	CRASSULACEÆ.	HOUSE-LEEKS.	*Sedum.*	Stone-crop.		14	400
52.	MELASTOMACEÆ.	MELASTOMES.	*Rheria.*	Deer Grass.		134	1250
53.	LYTHRACEÆ.	LOOSESTRIFES.	*Lythrum.*	Grass Poly.		30	300
54.	ONAGRACEÆ.	ONAGRADS.	*Œnothera.*	Evening Primrose.		22	450
55.	LOASACEÆ.	LOASADS.	*Mentzelia.*	Golden Bartonia.		10	70
57.	PASSIFLORACEÆ.	PASSIONWORTS.	*Passiflora.*	Passion Flower.		19	240
58.	CUCURBITACEÆ.	CUCURBITS.	*Sicyos.*	Wild Cucumber.		68	
60.	CACTACEÆ.	INDIAN FIGS.	*Opuntia.*	Indian Fig.		13	200
63.	UMBELLIFERÆ.	UMBELWORTS.	*Osmorhiza.*	Sweet Cicely.		152	1500
64.	ARALIACEÆ.	ARALIADS.	*Aralia.*	Pettymorrel.		38	200
65.	CORNACEÆ.	CORNELS.	*Cornus.*	Dogwood.		12	50

Cohort B. The Gamopelæ.

No.	Family	Common Name	Genus	Common	#	Count
66.	CAPRIFOLIACEÆ.	HONEYSUCKLES.	*Lonicera.*	Honeysuckle.	13	200
67.	RUBIACEÆ.	MADDERWORTS.	*Houstonia.*	Bluets.	337	4100
68.	VALERIANACEÆ.	VALERIANS.	*Valerianella.*	Lamb Lettuce.	9	175
69.	DIPSACEÆ.	TEASELWORTS.	*Dipsacus.*	Teasel.	5	150
70.	COMPOSITÆ.	ASTERWORTS.	*Erigeron.*	Robins' Plantain.	766	9000
71.	LOBELIACEÆ.	LOBELIADS.	*Lobelia.*	Cardinal Flower.	23	400
72.	CAMPANULACEÆ.	BELLWORTS.	*Campanula.*	Hare-bell.	30	600
73.	ERICACEÆ.	HEATHWORTS.	*Gaultheria.*	Checkerberry.	61	1330
74.	AQUIFOLIACEÆ.	HOLLYWORTS.	*Prinos.*	Black Alder.	4	
77.	EBENACEÆ.	EBENADS.	*Diospyros.*	Persimmon.	5	250
78.	SAPOTACEÆ.	SOAPWORTS.	*Bumelia.*	Bumelia.	23	320
81.	PRIMULACEÆ.	PRIMWORTS.	*Dodecatheon.*	American Cowslip.	21	300
82.	PLANTAGINACEÆ.	RIBWORTS.	*Plantago.*	Plantain.	3	200
83.	PLUMBAGINACEÆ.	LEADWORTS.	*Statice.*	Marsh Rosemary.	8	200
84.	LENTIBULACEÆ.	BUTTERWORTS.	*Pinguicula.*	Butterwort.	4	175
85.	OROBANCHACEÆ.	BROOMRAPES.	*Epiphegus,*	Beechdrops.	11	150
86.	BIGNONIACEÆ.	TRUMPET-FLOWERS.	*Catalpa.*	Catalpa.	53	450
88.	SCROPHULARIACEÆ.	FIGWORTS.	*Verbascum.*	Mullein.	157	1900
89.	ACANTHACEÆ.	ACANTHADS.	*Ruellia.*	Ruellia.	120	1350
90.	VERBENACEÆ.	VERVAINS.	*Verbena.*	Verbena.	59	700
91.	LABIATÆ.	LABIATE PLANTS.	*Salvia.*	Sage.	126	2600
92.	BORRAGINACEÆ.	BORRAGEWORTS.	*Myosotis.*	Forget-me-not.	68	1200
93.	HYDROPHYLLACEÆ.	HYDROPHILLS.	*Hydrophyllum.*	Water-leaf.	15	150
94.	POLEMONIACEÆ.	PHLOXWORTS.	*Phlox.*	Phlox.	8	150
95.	CONVOLVULACEÆ.	BINDWEEDS.	*Calystegia.*	Rutland Beauty.	32	800

				GENERA	SPECIES
96.	SOLANACEÆ.	NIGHTSHADES.	*Solanum.*	Potato.	66	1250
97.	GENTIANACEÆ.	GENTIANWORTS.	*Gentiana.*	Gentian.	49	520
98.	LOGANIACEÆ.	LOGANIADS.	*Gelsemium.*	Yellow Jessamine.	30	350
99.	APOCYNACEÆ.	DOG-BANES.	*Apocynum.*	Dog's-bane.	103	900
100.	ASCLEPIADACEÆ.	ASCLEPIADS.	*Asclepias.*	Milk-weed.	146	1300
101.	OLEACEÆ.	OLIVEWORTS.	*Fraxinus.*	Ash.	18	280

COHORT C. APETALÆ.

102.	ARISTOLOCHIACEÆ.	BIRTHWORTS.	*Asarum.*	Wild Ginger.	9	130
103.	NYCTAGINACEÆ.	MARVELWORTS.	*Mirabilis.*	Four-o'clock.	16	110
104.	POLYGONACEÆ.	SORRELWORTS.	*Polygonum.*	Knot-weed.	33	690
106.	CHENOPODIACEÆ.	GOOSE-FOOTS.	*Chenopodium.*	Pigweed.	78	530
107.	AMARANTACEÆ.	AMARANTHS.	*Amarantus.*	Prince's Feather.	46	480
108.	LAURACEÆ.	LAURELWORTS.	*Sassafras.*	Sassafras.	50	450
109.	LORANTHACEÆ.	LORANTHS.	*Phoradendron.*	Mistletoe.	25	400
111.	THYMELACEÆ.	DAPHNADS.	*Dirca.*	Leatherwood.	40	375
112.	ELEAGNACEÆ.	OLEASTERS.	*Eleagnus.*	Oleaster.	4	30
113.	EUPHORBIACEÆ.	SPURGEWORTS.	*Euphorbia.*	Spurge.	190	3200
114.	URTICACEÆ.	NETTLEWORTS.	*Ulmus.*	Elm.	65	1000
119.	EMPETRACEÆ.	CROWBERRIES.	*Empetrum.*	Crowberry.	3	4
120.	PLATANACEÆ.	SYCAMORES.	*Platanus.*	Plane-tree.	1	5
121.	JUGLANDACEÆ.	WALNUTS.	*Carya.*	Hickory.	4	27
122.	CUPULIFERÆ.	MASTWORTS.	*Quercus.*	Oak.	8	260
123.	BETULACEÆ.	BIRCHWORTS.	*Betula.*	Birch.	2	65
124.	MYRICACEÆ.	GALEWORTS.	*Comptonia.*	Sweet Fern.	3	20
125.	SALICACEÆ.	WILLOW-WORTS.	*Salix.*	Willow.	2	220

NATURAL ORDERS.

Cohort D. Conoideæ.

126. CYCADACEÆ.	CYCADS.	*Zamia.*	Florida Arrow-root.	7	46
127. CONIFERÆ.	CONIFERS.	*Juniperus.*	Red Cedar.	20	110
128. TAXACEÆ.	YEWS.	*Taxus Canadensis.*	Dwarf Yew.	2	50

Cohort E. Spadiciflora.

129. PALMACEÆ.	PALMS.	*Sabal.*	Palmetto.	73	400
130. ARACEÆ.	AROIDS.	*Orontium.*	Golden Club.	46	240
132. TYPHACEÆ.	TYPHADS.	*Typha.*	Cat-Tail.	2	13
133. NAIADACEÆ.	NAIADS.	*Potamogeton.*	Pond-Weed.	9	60

Cohort F. Florideæ.

134. ALISMACEÆ.	WATER PLANTAINS.	*Sagittaria.*	Arrow-Head.	9	70
137. ORCHIDACEÆ.	ORCHIDS.	*Cypripedium.*	Lady's Slipper.	394	3000
139. AMARYLLIDACEÆ.	AMARYLLIDS.	*Agave.*	American Aloe.	68	400
140. BROMELIACEÆ.	BROMELIADS.	*Tillandsia.*	Long Moss.	22	170
143. IRIDACEÆ.	IRIDS.	*Iris.*	Flower-de-Luce.	52	550
143. DIOSCOREACEÆ.	YAM-ROOTS.	*Dioscorea.*	Yam-Root.	7	150
144. SMILACEÆ.	SARSAPARILLAS.	*Smilax.*	Green-Brier.	2	120
146. TRILLIACEÆ.	TRILLIADS.	*Trillium.*	Wake-Robin.	4	30
147. LILIACEÆ.	LILYWORTS.	*Lilium.*	Lily.	147	1200
148. MELANTHACEÆ.	MELANTHS.	*Veratrum.*	False Hellebore.	30	130
149. PONTEDERIACEÆ.	PONTEDERIADS.	*Pontederia.*	Pickerel-Weed.	6	30
150. JUNCACEÆ.	RUSHES.	*Juncus.*	Rush.	15	200

290 NATURAL ORDERS.

				GENERA.	SPECIES.
151. COMMELYNACEÆ.	SPIDERWORTS.	*Tradescantia.*	Spider wort.	16	260
152. XYRIDACEÆ.	XYRIDS.	*Xyris.*	Yellow-eyed Grass.	5	70

COHORT G. GRAMINOIDEÆ.

154. CYPERACEÆ.	THE SEDGES.	*Cyperus.*	Galingale.	120	2000
155. GRAMINEÆ.	THE GRASSES.	*Agrostis.*	Bent Grass.	300	3800

PROVINCE THIRD. ACROGENS.

156. MARSILIACEÆ.	PEPPERWORTS.	*Isoetes.*	Quillwort.	6	20
157. LYCOPODIACEÆ.	CLUB MOSSES.	*Lycopodium.*	Ground Pine.	5	200
158. EQUISETACEÆ.	HORSETAILS.	*Equisetum.*	Scouring Rush.	1	10
159. FILICES.	FERNS.	*Struthiópteris.*	Ostrich Fern.	200	2000
160. MUSCI.	MOSSES.	*Polytrichum.*	Pigeon-wheat.	45	2000
161. HEPATICÆ.	LIVERWORTS.	*Jungermannia.*	Scale-moss.	65	700

PROVINCE FOURTH. THALLOGENS.

162. LICHENES.	LICHENS.	*Sticta pulmonaria.*	Lungs of Oak.	58	2400
163. FUNGI.	MUSHROOMS.	*Bovista gigantea.*	Puff Ball.	600	4000
164. ALGÆ.	SEA WEEDS.	*Chara vulgaris.*	Chara.	280	2000

NASTURTIAN.

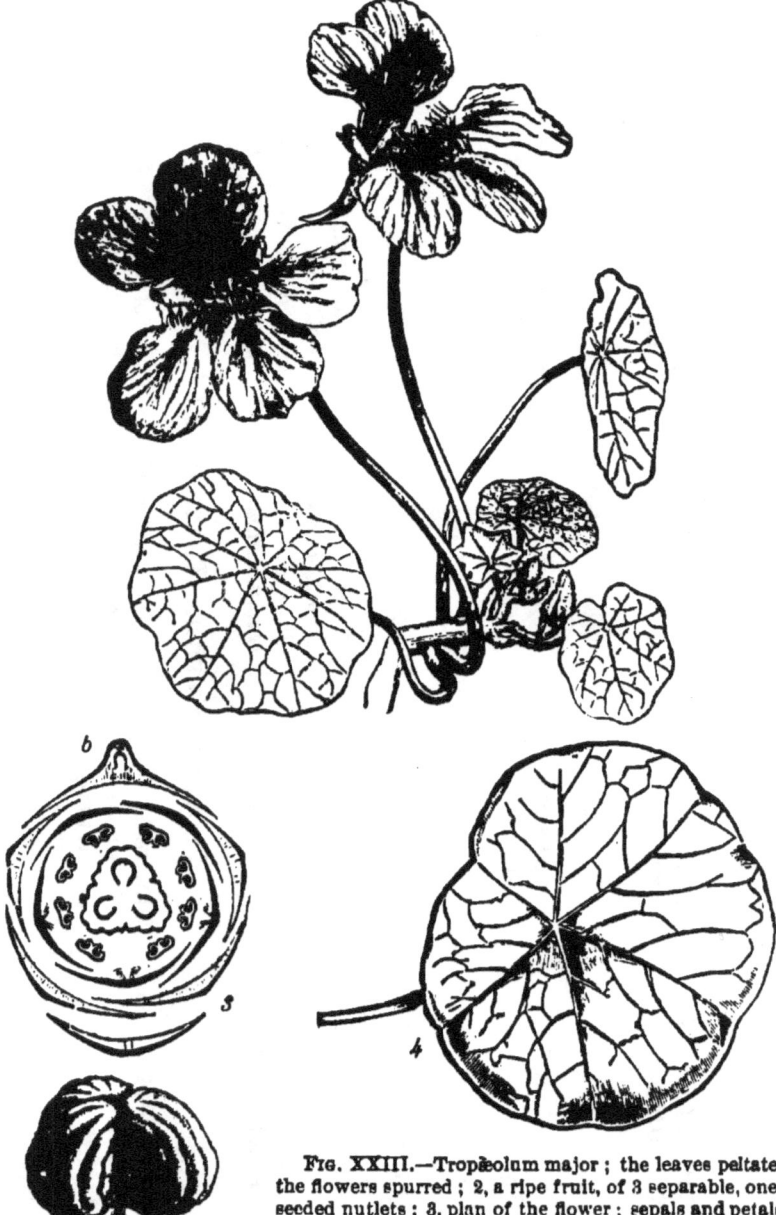

Fig. XXIII.—Tropæolum major; the leaves peltate, the flowers spurred; 2, a ripe fruit, of 3 separable, one-seeded nutlets; 3, plan of the flower; sepals and petals imbricated; stamens 8, carpels 3; *b*, the spurred sepal. See p. 95.

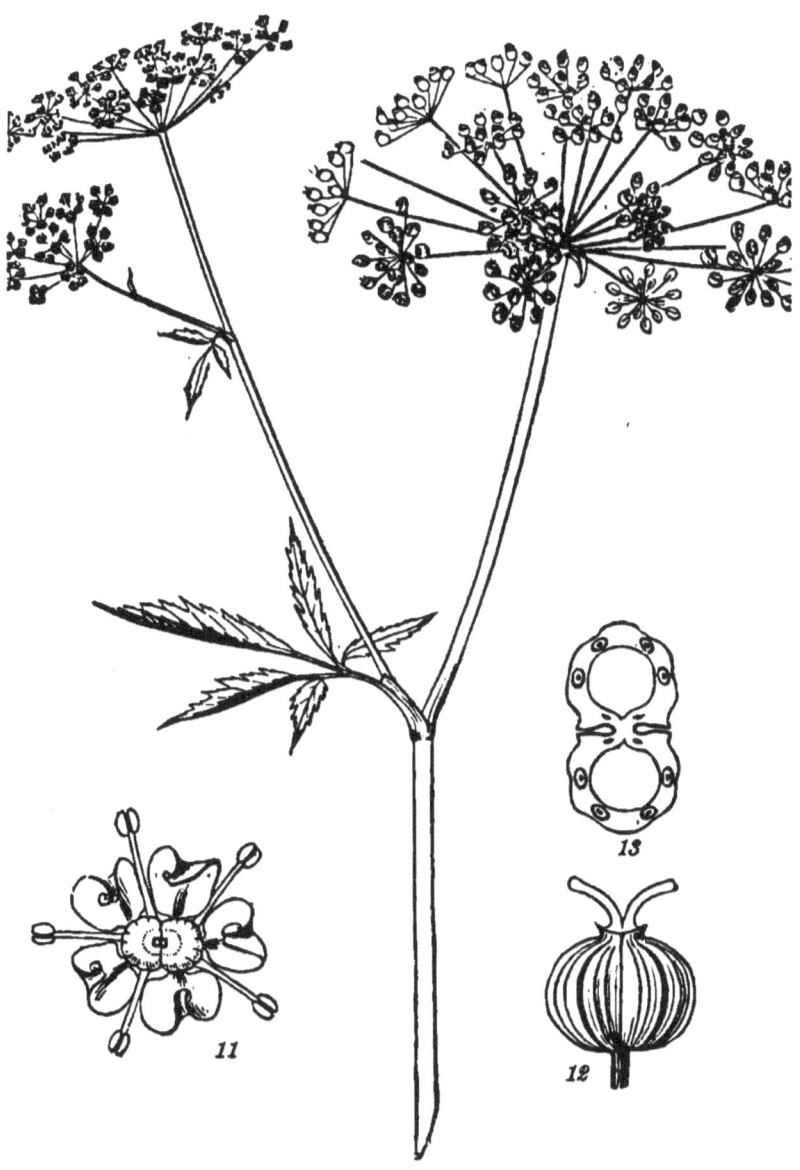

Fig. XXXIV.—10, Cicùta maculàta (Water Hemlock): 11, a flower enlarged; 12, a cremocarp; 13, cross-section of the same, showing the place of the 8 oil-tubes (vittæ). See p. 137.

ORGAN.	Life, Habit, Number, Place, Dehiscence, Kind, Construction, Form, Placentation, Size, Qualities, Appendages.
Plant, L.H.S.Q.	♃, herb erect, branching, 6 feet high, glabrous.
Root, L.K.	♃, of fibers, some of them thick, fleshy, oblong.
Stem, L.H.K.F.	Herbaceous, terete, hollow, striate or spotted with brown.
Leaves, L.P.C.F.S.Q.	Cauline, comp., on sheathing petioles; leaflets lanceolate.*
Inflorescence, P.K.A.	In terminal, compound umbels, involucre few-leaved.
Flower, N.C.	Numerous, complete, perfect, regular, 5-parted.
Calyx, F.Q.	Tube adherent to ovary, green, minute.
Sepals, L.N.P.F.	Minute teeth 5, epigynous, valvate in bud.
Corolla, F.Q.	Rotate, white.
Petals, L.N.P.F.	5, deciduous, epigynous, inflected at the point.
Stamens, N.P.C.	5, epigynous, diverging, complete.
Anther, D.C.F.	Opening lengthwise, introrse, oval.
Style, N.C.F.	2, short, distinct, slender.
Stigma, N.F.	2, club-shaped.
Ovary, C.F.Pn.	2-carpelled, 2-ovuled.
Fruit, N.D.K.F.Q.	A cremocarp, oval; carpels with 5 ribs and 4 vittæ.
Seed, N.C.F.Q.A.	1 in each carpel, suspended, albuminous.

LOCALITY.—*Swamps, Worcester, Mass.* (Date), *July, 1870.*

CLASSIFICATION.—POLYPETALOUS EXOGENS.

—UMBELLIFERÆ, THE UMBELWORTS.

NAME.—Latin, **Cicuta maculata.**

—English, *Spotted Water-hemlock.*

REMARKS.—*The veinlets terminate in the notches between the teeth. The herbage is said to be poisonous.*

ANTENNARIA.

The Record.—This plant, and the Order which it represents, offers so many peculiarities of inflorescence that a new tablet becomes necessary. (See Plant Record, Asterwort.)

ORGAN.	Life, Habit, Number, Place, Kind, Construction, Form, Size, Qualities of color, etc., Appendages.
Plant, L.H.S.Q.	♃, herb, in dry pastures, 6—9', woolly-canescent.
Stem, L.H.K.F.	Herbaceous, erect, simple, with runners at base of caulis.
Leaves, L.P.C.F.S.Q.	Decid., alternate, entire, obovate, oval-spatulate, and linear-oblong, pinni-veined, petiolate, exstipulate.
Petiole, F.S.Q.	Margined, 2'—1'—0', upper leaves sessile.
Inflorescence, P.K.	Terminal, in heads; heads clustered.
Head, K.F.S.	Dioecious, discoid, 3" diameter.
Involucre, K.F.	Imbricated, oval or hemispherical.
Scales, N.P.F.Q.	∞, appressed, ovate, scarious, white, ♂ obtuse, ♀ acute.
Receptacle, F.Q.	Flattish, naked.
Pales, N.P.F.Q.	None.
Ray flowers, N.K.F.Q.	None.
Disk flowers, N.K.F.Q.	$\sqrt{5/}$, ♂ and ♀ on different plants, tubular, 5-toothed, white.
Pappus, L.N.C.F.Q.	Persistent, 20, simple, capillary, white.
Stigmas, N.P.C.F.Q.	2, exserted, recurved, ♂ united, yellow.
Achenium, F.Q.	Linear, teretish, brown.
Embryo, P.F.	Axial, straight.

LOCALITY.—*Dayton, O.* (Date), *April 12.*

CLASSIFICATION.—GAMOPETALOUS EXOGENS.
  —Order, COMPOSITÆ, or THE ASTERWORTS.

NAME.—Latin, **Antennaria plantaginifolia**.
  —English, *Mouse-ear Everlasting.*

REMARKS.—*The pappus of the sterile florets consists of club-shaped knobby bristles poorly adapted to flying.*

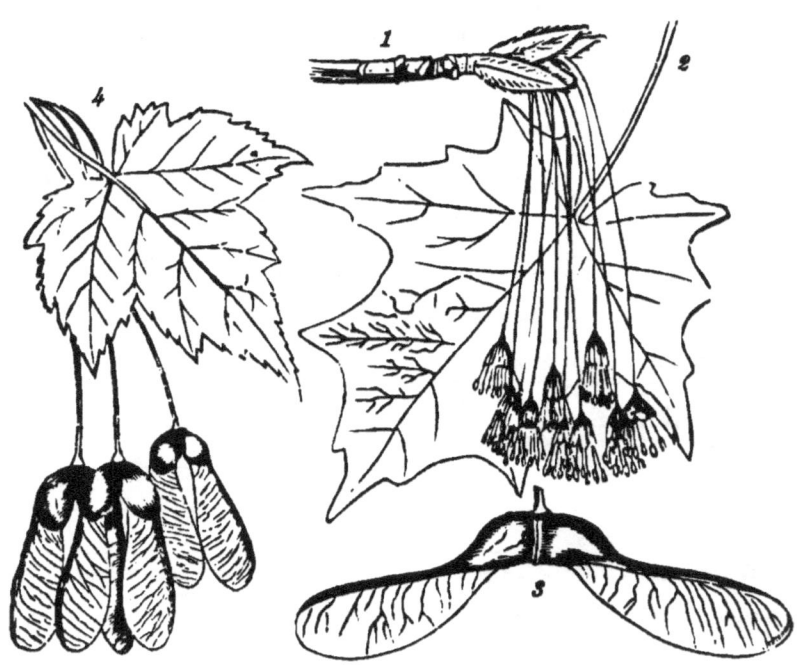

Fig. LI.—Flowers, leaf, and double samara of Acer saccharinum (Sugar Maple): 4, leaf and samaræ of Acer rubrum (Red Maple). See p. 188.

Fig. LVI.—Quercus: 1, leaf of Q. virens, Live Oak; 2, Q. Phellos, Willow Oak; 3, Q. imbricària, Shingle Oak; 4, Q. aquática, Water Oak; 5, Q. nigra, Black Jack; 6, Q. triloba, Downy-Black Jack; 7, Q. ilicifòlia, Bear Oak; 8, Q. rubra, Red Oak; 9, Q. palústris, Pin Oak; 10, Q. coccínia, Scarlet Oak; 11, Q. falcàta, Spanish Oak; 12, Q. alba, White Oak; 13, Q. obtusiloba, Iron Oak; 14, Q. macrocárpa, Mossy-cup Oak; 15, Q. bícolor, Swamp-White Oak; 16, Q. Prinos, Swamp-Chestnut Oak.

QUESTIONS.

I. ADAPTED TO THE FLOWERLESS PLANTS.

1. Distinguish two regions.
2. What parts are distinguishable in each region?
3. What is the form of the Root?
4. What purposes does it serve?
5. In which region is the Stem?
6. Describe its form, attitude, height or length.
7. Its habit as to branches.—Its kind as to scales or leaves.
8. Where are the Leaves placed? How arranged?
9. Shall we call them fronds? Why, or why not?
10. Are they simple, or compound?
11. What are their members? Have they veins?
12. Name their three kinds of veins.
13. What is the kind and mode of venation?
14. Define carefully the form of outline.
15. What is their quality of surface, or clothing?
16. What do you call their stalks, if any?
17. Where is the fruit produced?
18. What supports it? Describe the pedicel, if any.
19. Tell how the capsules open.
20. Point out the operculum, or the elastic ring.
21. Describe the peristome, if any.
22. What do the capsules contain?
23. What becomes of the spores? What is their use?
24. What if no more spores were produced?
25. Do you find any flowers? Of what size and appearance?
26. In what *subkingdom* is this plant classed?

27. In what sense is this a "Flowerless Plant"?
28. What is its order?—genus?—species?*
29. What is its popular name?

II. ON THE LEAF REGION OF A FLOWERING PLANT.

§ 1. The Plant.

Life. Is it an annual, a biennial, or a perennial?
Habit. Is it an herb, a shrub, vine, or tree?
    Describe its locality.
Size. What are its height, or length, and other dimensions?
Qualities. In surface, is it smooth, or rough, or hairy?
    What term defines its color or special hue?

§ 2. The Root.

Life State its term of duration.
Form. Is it axial, or inaxial? Of what special form?

§ 3. The Stem.

Life. What is its duration and substance?
Habit. In growth, is it exogenous, or endogenous?
    What is its direction or posture?
    What is its habit of branching?
Kind. Is it scaly or leafy?—under or above ground?
    Is it a caulis, trunk, bulb, or rhizome, etc.?
Form. Is it solid, or tubular, angular, or terete, etc.?

§ 4. The Leaves.

Life. Are the leaves deciduous, or evergreen?
Place. How are they folded in vernation?
    What is their position on the plant?
    How are they arranged among themselves?
Construction. Describe their veins and venation.
    Of what numbers are they constituted?
    Are they simple, or compound?
    Describe the mode of composition.
Form of blade. What term or terms define their outline?

---

* The scientific name of a plant, or its genus and species, if not communicated, may be determined, after analysis, by the aid of a Descriptive Flora with analytical tables.

What term defines the apex?—or base?
The margin—is it dentate?—serrate?—entire?—etc.
Size. State their measurements.
Quality. Describe their surface-quality, or clothing.

§ 5. The Petiole.

Form, etc. State the form, size, and quality of the petiole.

§ 6. The Stipules.

Life, etc. State their duration, kind, and form.

III. ON THE FLOWER REGION OF ANY FLOWERING PLANT
(EXCEPT THE COMPOSITES, SEDGES AND GRASSES).

§ 1. Inflorescence.

Place. Define the position and posture of the flowers.
Kind. In general, is the inflorescence solitary, or centripetal? etc.
In particular, is it a raceme?—spike?—spadix?
Appendages. Name the flower-stalks, if any.
Describe the bracts, if any.
The involucre—the involucels.
Point out the scales—the pales.
Point out the rachis—the torus.

§ 2. The Flower.

Number. What is the radical number of the flower?
Construction. Name all its organs. How many are there of each?
Is it complete? What is lacking?
Is it regular? How irregular?
Is it symmetrical? How unsymmetrical?
Why polypetalous, or gamopetalous?
Distinguish the torus—the disk.

§ 3. The Calyx and Corolla, or Perianth.

Form. Is it polyphyllous? Are its leaves united?
What term defines the special form?
Quality. In surface, is it smooth?—hairy?—granular?
What term or terms define the color?

§ 4. Sepals and Petals.

Life. How long is their duration? What term defines it?
Number. How many are there?
Place. Define carefully their æstivation.
Form. Describe the pattern of their outline.
Nectary. Describe it, if conspicuous.
Lip. What is the special form of the lip, if any?
Corona. Describe its situation, parts, and form.

§ 5. The Stamens.

Number. How many?—definite, or indefinite?
Place, as to the adjacent organs—are they opposing? alternating? etc.
    Are they exserted?—included?—connivent?
    Are they ascending?—declining?
    Why are they hypogynous? Why epipetalous? etc.
    How connected—Gynandrous? Syngenecious?
    Diadelphous?—Monadelphous? etc.
Construction. Of what members are they composed?
    What member is lacking when sessile?—sterile?
    Are they didynamous?—tetradynamous?

§ 6. Anther.

Place. How is it attached to the filament? Which way does it face?
Dehiscence. How does it open? In what direction?
Construction. In how many cells is it divided? What appendages are there?
Form. What term defines their shape?

§ 7. Pollen, Pollinia.

Form. Describe the pollen, or pollinia, as to form and quality.

§ 8. The Style.

Number. How many styles? Are they united or separate?
Place. How situated on the ovary? In what posture?
Form. What term describes their form?

§ 9. The Stigma.

Number. How many stigmas?
Place. How attached to the style? When is the stigma sessile?
Form. Of its many shapes, which is this?

§ 10. Ovary.

Construction. Is it simple? How compounded?
How many cells? Is it free or adherent?
Form. What term or terms define its shape?
Placentation. Is the placenta central, free-central, or parietal?

§ 11. Ovules.

Place. What is their position in the cell?
Construction. Are they anatropous?—orthotropous?

§ 12. Fruit.

Number. How many carpels, whether distinct or united?
How many cells has the fruit? Same as the ovary?
Dehiscence. Is this a dehiscent, or indehiscent fruit?
What term describes its mode of dehiscence?
Kinds. Of the twenty-seven special kinds, which is this?
Analyze its coverings, substance, cells, valves, carpophore, etc.
Form. What term or terms indicates its shape?
Quality. Describe its color, texture, and clothing.

§ 13. Seed.

Number. Are there few, or indefinite, or how many?
Construction, in regard to albumen, and cotyledons?
Form. What term defines their shape?
Quality, as to color, surface, or clothing.
Appendages. Have they wings?—or a coma?—or an aril? etc.

IV. ON THE FLOWER REGION OF THE COMPOSITES.

§ 1. Head.

Kind. Are the heads diœcious?—or monœcious? etc.
Form. Are they discoid?—radiate?—radiant?

§ 2. Involucre.

Kind. Is the involucre simple?—imbricated?—calyculate?
Form. What term defines the shape?

§ 3. Scales and Pales.

Number. Few?—definite?—or indefinite?

Place. Are they erect?—appressed?—loose? etc.
Form. Define their outline, margin, apex.
Qualities. What texture?—surface?—color?

§ 4. Receptacle.

Form. Is the receptacle flat?—convex?—conical?
Quality. Is it naked?—chaffy?—bristly?—alveolate?

§ 5. Rays, or Ray Flowers.

Number. How many?—in how many rows?
Kind. Are they perfect?— ☉?— ♀?—sterile?—fertile?
Form. Are they ligulate?—tubular?—linear?—oblong? etc.
Quality. What is their color?

§ 6. Disk Flowers.

Number. Can they be easily counted?
Kind. Are they perfect?— ☉?— ♀?—fertile or sterile?
Form. Are they tubular?—4-toothed?—5-toothed?
Quality. Of what color?

§ 7. Stigmas.

Place. Are they straight?—incurved?—recurved?
Form. Are they flat?—terete?—pointed?—truncate?
Quality. How is their surface clothed?

§ 8. Fruit (—achenia or cypsela).

Form. Describe its shape, whether linear, terete, compressed, etc.
Quality. What of its surface, and color?

§ 9. Pappus.

Life. Is the pappus deciduous or persistent on the fruit?
Number. Of how many bristles or scales.
Construction. Is it stiped? or simple? or double?
Form. Is it capillary? scaly? plumous? barbed?
Quality. What is the color?

V. ON THE FLOWER REGION OF THE SEDGES AND GRASSES.

§ 1. Inflorescence (general).

Place. Are the flowers axillary, or terminal, or both?

Kind. Are they in panicles, or spikes?—compact or loose?
Appendages. How are the branches of the panicle arranged?

§ 2. {Spikes (of the Sedges) or Spikelets (of the Grasses).

Number. Are there few, or many? How many?
Kind. Are their flowers ♀, or ♂, or perfect (☿)?
 Are they monœcious, or diœcious?
Form. Describe their particular shape, as terete, ovoid, etc.
Size. How long, and large are they?
Quality. Describe their color or surface character.

§ 3. Flowers.

Number in the spike or spikelet—how many?
Kind. Are they fertile? or sterile? or both?
 How many stamens have they—1, 2, 3, or more?
 How many stigmas—1, or 2?

§ 4. Glumes.

Number. Are there 1, or 2, glumes, or none?
Place. Are they arranged in 2 rows? or 3? or imbricated all around?
Form. What is their shape—as lanceolate, carinate, bifid?
Size. How long are they relatively?—intrinsically?
Qualities. In texture, herbaceous? scarious? What in color?

§ 5. Pales, or Perianth.

Kinds. Are there setæ? or petals? or pales? or a perigynium?
Number. How many pieces in each flower?
Place. What is their posture—erect? recurved? free?
Form. What the special form of the perigynium?
Size. What is its relative length?
Quality. What of its surface?—its color?

§ 6. Awns.

Place. Are they on the back or the tip of the glume or pale or rudiment?
Form. Are they straight, or bent?—bristle-form, or gossamer?
Size. What is its relative length?
Quality. Are they scabrous, or smooth, or feathery?

§ 8. Rudiments (abortive flowers).

Number. Are there 1, or 2, or several?
Kind. Are they staminate, or neutral?
Form. Are they a pale?—a pedicel? or an awn?
Quality. Are they ciliate? scabrous? hairy?

§ 9. Anthers.

Place. How are they attached to the filament?

§ 10. Stigmas.

Form, etc. Are they plumous?—curved?—erect?

§ 11. Grain.

Kind. Is it an achenium, a cariopsis, or a utricle?
Form. Is it triquitrous? lens-shaped? oblong? etc.
Appendages. Is it tipped with a tubercle?

VI. ON THE ACTION, ETC., OF PLANTS.

§ 1. Fertilization.

1. Does this plant produce any nectar?
2. Describe the place and form of the nectaries.
3. What seems to be the primary use of them?
4. Is the plant wind-fertilized, or insect fertilized?
5. What arrangement, if any, prevents self-fertilization?
6. Is the flower proterandrous? Is it protogynous?
7. Is it dimorphous? How does this appear?
8. How does this favor crossing?
9. Do the stamens show irritability? How do they act?
10. Is the pollen in grains, or in pollinia?
11. Describe the pollinia. How and by what extracted?
12. How are they brought to bear on the stigma?
13. What arrangement to save the honey from the ants?
14. How is it secured against rain and dew?
15. Are the flowers, any or all, cleistogene? Describe such.
16. How is the pollen thus economized?

§ 2. Sleep.

1. Is this plant sensitive to night and day?—light and shade?
2. What are the indications of it?

3. At what hour does it open its flowers?
4. How long do they remain open? When do they close?
5. Do they open and close more than once?
6. Do the leaves change position at night, or in shade?
7. Describe the change.
8. How is the nectar affected by the flowers' closing?

§ 3. Irritability.

1. Is this plant sensitive to touch? In which organ?
2. How is this indicated?

§ 4. Movements.

1. How is this vine furnished for climbing?
2. Has it any special organs? What are they?
3. Have you observed their movements? Describe them.
4. What seems to be the purpose of this motion?
5. After it has reached an object, how does the tendril act?
6. How does a vine without tendrils climb?
7. Is it aided by its petioles? Explain.
8. If it be a twining vine, which way does it turn?
9. To find a support, what movement have you observed?
10. Is it aided by hairs, or prickles?—by rootlets?

§ 5. Classification.

1. Is this plant an Exogen, or an Endogen?
2. By what marks do you determine this?
3. Has it a stigma, and a seed-vessel? Is it then a Gymnosperm?
4. Have you determined its Order?—its Genus?—its Species?

INDEX.

PRONOUNCING, GLOSSARIAL AND REFERENTIAL.

A.

A (in composition) prefixed to a Greek word, signifies *without*; as *apetalous*, without petals.
Abies, 218.
Abórtive, not developed, imperfect.
Abórtion, non-development of a part.
Abrupt at base, truncate.
Absinthe, 147.
Acàcia, a-kà-shi-a, 134.
Acaulescent, apparently stemless, 54, 66, 143.
Accessory, something superadded.
Accrescent, growing after flowering; sc. calyx.
Accumbent, lying against the edge, 103. See Cotyledons.
Acer, 192.
Acerous, needle-shaped, 214.
Achenium (a-kén-i-um), plu. achenia, 48, 147, 178.
Aconite, 64.
Acorns, 209, 210.
Acùleate, armed with prickles.
Acùminate, extended into a point.
Acute, ending in a sharp angle.
Adam-and-Eve (Apléctrum), 236.
Adherent, growing to, 87.
Adherent Ovary. Current. 110, 233.
Adiántum, 27.
Adnate, growing fast to, 74, 184.
Adnate stipules, Rose.
Adònis, 64.
Adventitious, out of the normal position, 129.
Aerial region, 20.
Æsculus, 194.
Æstivàtion, 42.
Affinity, resemblance in essential organs
African Hemp, 258.
Agapanthus, 258.
Agàve, a-gà-ve, 263.
Aggregate, assembled close together.
Aglumaceous, without glumes.
Air plants, 239.
Ala, pl. Alæ, wings, 118.

Albùmen, 33, 42, 111.
Albùminous, 186.
Albùrnum, the sap-wood (p. 107).
Alfillirèa or Alfilaria, the "Piu Grass" of the Pacific Coast (Erodium cicutarium), 87.
Algæ, seaweeds, 27.
Allium, 258.
Aloe, American, 263.
Aloes, 258.
Alpine Primrose, 168.
Alternate generation, 22.
Alternate leaves, 193.
Alternating stamens, 33.
Alveolate, with pits, as a honeycomb.
Alyssum, 108.
Amaryllis, 263.
Amaryllidàceæ, 262.
Ament, a deciduous spike, 208.
Amórphous, without definite fo m.
Amplexicaul, stem-clasping, 96, 251.
Amygdalus, 116.
Anagàllis, 168.
Analysis, 13.
Anátropous ovule, 56.
Ancipital, two-edged, 244.
Androgynous, staminate and pistillate flowers together in a cluster, 266.
Anemòne, 61.
Angiosperms, 220.
Anise, 136.
Annual, living one year; yearly, 68.
Annular cells, cells distended with rings, 225.
Antennària, 189.
Anterior, facing outward.
Anthèmis, 147.
Anthelmíntic, expelling or killing worms.
Anther, innate, attached by base.
Anther, adnate, attached by back.
Anther, versatile, attached by middle.
Anther, valvate, dehiscing by valves.
Anthèsis, the act of flowering.
Antheridia, staminate organs of Mosses, etc., 14.

INDEX. 307

Antirrhìnum, 176.
Apétalous, without petals, 56.
Aphýllous, without leaves.
Apópysis, a swelling. See Fig. I, 7, a.
Apex, the summit or tip, 21.
Apple Tree, 107, 111.
Apple Moss, 17.
Appréssed, closely applied, but not adhering to; the same as adpressed.
Apterous, without wings.
Aquatic, living in the water.
Araceæ, 232.
Arbóreus, arborescent, tree-like.
Arbùtus, trailing, 158.
Arcuate, arched or curved like a bow.
Arctostáphylus, 158.
Aril, an extra seed-envelope, as in Enonymus.
Arisæma, 231.
Aristate, awned; bearing an awn.
Armed, bearing prickles, spines, etc.
Arnica, 147.
Aroids, 232.
Articulated, jointed.
Artemísia, 147.
Arum, 232.
Ascending, arising obliquely; assurgent.
Ascidia, leaves holding water, 161.
Asclepiadàceæ, 197.
Asclèpias, 197, 199.
Ash, 93.
Aspídium, 27.
Assafœtida, 137.
Assimilate, to digest as food.
Aster, China, 147.
Asterworts, 146.
Attar of Roses, 116.
Arctostáphylus, 158.
Auricula, 168.
Auriculate, with ear-shaped lobes, 74.
Awn, the beard of Barley, and the like, 276.
Axial root, 50.
Axil, the angle between the petiole and branch, on the upper side, 90.
Axillary, growing out of the axils, 90.
Axis, the central body or column, 13.

B.

Baccate, berry-like.
Balm.
Balsamine, 95.
Bamboo, 225, 280.
Banner, same as vexillum, 118.
Bark, the outer layers of Exogenous stems, 107.
Barley, 279.
Bartrámia, 18.
Bartram, John, 18.
Básilar style, attached to the base of the ovary. Brunella, 180.
Bath Flower, 249.
Bdéllium, 137.
Beaked, ending in an extended tip, 267.
Bean, 125.
Beárberry. See Arctostaphylus.
Bearded, with awns, or tufted hairs.

Beech, 111. Beech Tree, 213.
Begònia, 120.
Bellworts, 251.
Berry, a fruit with its seeds immersed in [pulp.
Bi, Bis (in compound words), twice.
Biennial, of 2 years, 125.
Biennial-fruit, 210.
Bifid, cleft into 2 parts, 79.
Bifoliate, with two leaflets.
Big Trees of Calavèras, 221.
Bilabiate, two-lipped, 173.
Binate, two growing together. See Bifoliate.
Bindweeds, 187.
Bipinnate, twice pinnated.
Bipinnatifid, twice pinnatifid, 24.
Bird Knotweed, 201.
Biternate, twice ternate, 58.
Bivalved, two-valved.
Blackberry, 116.
Blade, the main part of a leaf.
Blinding Tree, 207.
Blanched, whitened for want of light; the same as etiolated.
Bloodroot, 64.
Bloom, a fine white powder on some plants.
Blueberry, 158.
Blue Curls, 179.—Flag, 241.—Grass, 274.—Violet, 77.
Blue-eyed Grass, 244.
Bouncing Bet, 83.
Boxberry, 147.
Boxwood, 207.
Brachiate, with opposite branches.
Bract, a reduced leaf near the flowers, 56.
Brácteoles, Bractlets, reduced bracts.
Brake, Common, 27.
Branches, the divisions of a stem.
Brassica, 103.
Brazil Wood, 125.
Bristles, stiff, sharp hairs.
Brunélla, 180.
Bryólogy, the science of Mosses.
Bud, The, a rudiment, 186.
Bud-scales, reduced leaves covering the bud.
Bulb, an underground bud, 29, 35.
Bulblets, little bulbs formed in the leaf-axils and falling off.
Bulbous Crowfoot, 50.
Bulb, a scaly, 35.
Bulb, a coated, or tunicated, 35.
Bulrushes, 270.
Burgundy Pitch, 223.
Butter-and-Eggs, 173.
Buttercup, 48.
Butterfly Weed, 199.
Buxus, 207.

C.

Cabbage, 103.
Cabbage Palmetto, 224.
Cadùcous, dropping off early, 66.
Cæspitous, forming tufts, or a turf.
Calàdium, 233.
Cálamus Rudéntum, 227.
Calceolària, 176.
Calico Bush, 155.

INDEX.

California Poppy, 71.
Calla, 232.
Calopogon, 235.
Calyculate, having an outer calyx, or calyx-like involucre.
Calyptra, the cap of a Moss-capsule, 14.
Calyx, the outer floral envelope, 31.
Calyx free, not joined to other organs, 110.
Calyx inferior, the same as calyx free, 108.
Cambium, the new layer of wood, next under the bark.
Ca-měl'-li-a, 115.
Cámomile, 147.
Campánulate, bell-shaped, 181.
Camwood, 125.
Canada Balsam, 223.
Candy Tuft, 103.
Canéscent, whitish with minute hairs.
Capers, 97, 207.
Cápillary, very fine, hair-like, 167.
Cápitate, inflorescence head-shaped.
Capsella, 100.
Capsule, a dry, dehiscent fruit, 14, 33.
Capsular, of or like a capsule.
Caraway, 136.
Cárdamine (car-da-mì-ne), 102.
Carex, 265.
Carinate, boat-shaped, keeled, 118.
Carinæ, the 2 lower petals of a papilionaceous flower, 118.
Carnivorous Plants, 161.
Carob, 125.
Carpels, the divisions of a fruit, 48, 220.
Carpels distinct. Thalictrum.
Carpinus, 214.
Cárpophore, the fruit-bearer, 85, 133.
Carrion Flower, 199.
Carrot, 136.
Cárthamus, 147, 246.
Carum, 134.
Cáruncle, an appendage of a seed, 205.
Caryophyllaceæ, 82.
Caryophyllaceous flower, i. e., 5 petals in a tubular calyx.
Caryópsis, a fruit like a Wheat-kernel, with the seed inseparable from its coat, 274.
Cascarilla, 207.
Cassava, 207.
Cassias, 123.
Castor oil, 206.
Catch Fly, 83.
Cátechu, 125.
Catkin, the same as ament, 208.
Catmint, 178.
Caudex, the stem of a Palm, etc., 80, 224.
Cauléscent, having a stem above-ground, 54.
Cauline, of the stem.
Caulis, an herbaceous stem, 29.
Cedars, Giant, 221.
Cedars, Red, 223.
Celery, 136.
Cellular tissue, 17, 225.
Centrifugal inflorescence, 104, 140.
Centripetal inflorescence, 41, 126.

Century Plant, 263.
Cereal, relating to grains, corn, etc.
Cérnuous, nodding (less than pendulous).
Chaff, the same as pales, 139, 273.
Chalaza, the place where the ovule joins its stalk.
Chamærops, 226.
Chartáceous, with texture like paper.
Checkerberry, 147.
Cheiránthus, 103.
Chelidònium, 71.
Cherry, 116.
Cherry Laurel, 116.
Chervil, 136.
Chestnut, 210, 213, 214.
Chickweed, 78.
Chick Wintergreen, 164.
Chickory, 147.
Chimáphila, 153, 154.
China Aster, 147. [136.
Chlorophyl, the green grains in leaf-cells,
Chrysánthemum, 147.
Cicely, 131.
Cichòrium, 147.
Cicuta, 137
Cilia, plu. ciliæ, hairs like the eyelash, 14, 18.
Ciliate, fringed with hairs, 132.
Cinèreous, ash-colored, ash-gray.
Cinnamon Fern, 26.
Cinquefoil, 112.
Circinate, rolled inward from the top, 26.
Circulation of the sap, 110.
Cirrhous, furnished with a tendril, 117.
Circumscíssile, opening by a lid, all around, 48; Fig. XLIII, 5.
Clarkia, 131.
Clavate, club-shaped.
Claw. See Unguiculate, 85, 110.
Claytonia, 43.
Clayton's Osmunda, 24.
Cleistogene flowers, never opening, 74.
Clématis, 64.
Cliánthus, 125.
Climbing Plants, 186.
Climbing Fern, 27.
Clintonia, 256.
Clove Pink, 80.
Clover, 125.
Club Mosses, 27.
Cóchleate, spiral like the snail-shell.
Còcoanut, 226.
Còcoanut Palm, 228.
Coherent, united as to similar parts.
Cohesion, union of similar parts.
Colocásia, 233.
Collateral, placed side by side.
Cólumbine, 64.
Colored, of any color except green.
Column, combined stamens and styles, Cypripèdium, 235.
Coma, the long hairs of a seed, as cotton, etc., 197.
Commissure, the joined faces of the carpels of a cremocarp, 135.
Complete Flower, having the 4 kinds of organs, 47.

INDEX. 309

Compound leaf, having several leaflets, 58.
Compósitæ, 146.
Conduplicate, leaf folded, the two halves face to face.
Cone, the scaly fruit of the Pines, etc., 215. [ent.
Cónfluent, uniting; same as coher-
Coníferæ, 220.
Conium, 137.
Cónjugate, united by pairs.
Connate, growing together, as leaves, etc.
Contorted, twisted; petals over-lapping all one way, 183.
Connéctile, that part of the filament which connects the two anther cells, 180.
Connivent, converging toward each other.
Convallària, 256.
Converging petals, see Connivent, 151.
Convolute, see Contorted. Also when the leaves or petals are rolled one within another.
Convolvulaceæ, Convolvulus, 187.
Copaiva balsam, 125.
Cordate (leaf), heart-shaped, 66, 73.
Coreópsis, 147.
Coriaceous, leather-like, 54.
Córiander, 136. [258.
Corm, a solid bulb-like stem, 50, 231.
Corn Cockle, 83,—Speedwell, 170.
Corólla, the inner floral envelope, 32.
Coròna, a crown in the midst of the flower, 196, 262.
Corymb, a level-topped cluster, centripetal, 114, 155.
Corymbous, of or like a corymb.
Costate, with rib-like ridges,
Cotton, 104.
Cotton Grass, 270.
Cotyledon (seed-lobe), 42, 86, 110, 186.
Cotyledon accumbent, 102. [104.
Cotyledon incumbent,
Cotyledon conduplicate, Mustard.
Cowslip, 161, 168.
Cow Tree, 199.
Crab Tree, 117.
Cranberry, 158.
Cranesbill, 83.
Creeper, a prostrate stem under or above ground, [102.
Crémocarp, the fruit of the Umbelworts,
Crenate, with rounded teeth, 11, 177.
Crénulate, the rounded teeth small.
Cress, Toothroot, 101.
Crest, an elevated ridge.
Crinum, 263.
Cristate, having an elevated ridge.
Crocus, 246.
Croton Oil, 206.
Crowfoot, 46, 62.
Crown of the root, 54.
Crown Imperial, 238.

Cruciferæ, 103. [100.
Cruciform corolla, cross-shaped.
Cryptogàmia, 16.
Cryptogams, 27.
Cryp-tóg-a-mious, 16, 27.
Cucullate, hood-shaped, 73.
Cùcumber, Indian, 249.
Cucumber, Squirting, 95.
Culm, the straw of the grasses, 264.
Cummin, 136.
Cuneate (leaf), cuneiform, wedge-shaped.
Cupulíferæ, 213.
Cúspidate, with a sharp, slender point. [ing.
Cuticle, the outer skin or cover-
Cyclamen, 168.
Cydònia, 117. [104.
Cyme, a centrifugal cluster,
Cyperàceæ, 268.
Cýperus, 264, 268.
Cypripedium, 238.
Cypsela, the fruit of the Compósitæ, 142.

D.

Daffodil, 261, 268.
Dahlia, 147.
Dalbergia, 125.
Damask Rose, 115.
Dandelion, 95, 143, 177.
Darlingtonia, 160.
Date Palm, 225, 228. [son, 33.
Deciduous, falling at the end of the sea-
Dehíscence, act or manner of opening.
Decompound, much compounded, 129.
Decurrent leaves, running down the stem.
Decumbent, first erect, then prostrate.
Definite, of a special number.
Defoliàtion, casting off of leaves.
Deltoid, form of the Gr. letter Δ. [form.
Dendroid, tree-like in
Dentària, 102.
Dentate, with teeth turned outward.
Depauperate, less developed than usual.
Depressed, flattened from above.
Desmodium gyrans, 124.
Di (in Gr. compounds), two.
Diadélphous, stamens in two sets, 118.
Diagnósis, the distinctive character.
Diándrous, having two stamens.
Diánthus, 82, 83.
Dichótomous, forked or 2-cleft, 17, 78.
Di-cót-y-léd-o-nous, embryo 2-lobed, 66.
Didynamous, with 2 long and 2 short stamens, 174.
Diffenbáchia, 233. [ing.
Diffuse, much branched and spread-
Digitate, leaflets distinct, palmately arranged, 194.
Digitalis, 176.
Dill, 136.
Dímerous, flowers two-parted. Circæa, 131.
Dimórphism, 162, 203.

310 INDEX.

Diœcious, staminate and pistillate flowers on different plants, 137.
Dionæa, 160.
Dipterix, 125.
Dipterous, with two wings.
Dischidia, 199.
Discoid Head, 145.
Disk, a layer between the stamens and ovary, Alchemilla. 184.
Dissected, cut into deep lobes, incised.
Dis-ti-chous, arranged in two rows.
Divaricate, wide-spread, straggling, 24.
Divergent, spread'g apart, more or less.
Dock, 50, 208.
Do-de-cáth-e-on, 162, 168, 177.
Dogtooth Violet, 29.
Dorsal, on the back.
Double Rose, 114.
Double Pink, 82.
Douglas Fir, 221.
Downy, clothed with short, weak hairs.
Dracèna, 258.
Dragon's Blood, 258.
Dragon's Root, 231.
Drósera, 161.
Drupe, a stone-fruit, as Cherry, Hickory.
Dumb Cane, 233.
Durâmen, heartwood, 107.

E.

E, or Ex (in composition), without; as Ebracteate, without bracts.
Elecampàne, 147.
Elliptical, form of an ellipse.
Elm, 36, 176.
Elóngated, lengthened, extended.
Emárginate, notched at the end, 133.
Embryo, straight; convolute, 32, 40.
Embryo coiled around albumen.
Enchanter's Nightshade, 131.
Endogens, 33, 229.
Endogenous structure, 225.
Ensiform, sword-shaped, 241.
Entire Margin, even-edged, 31.
Ephemeral, enduring for one day.
Epi (in composition), upon, as
Epidermis, same as cuticle.
Epigèa 158.
Epigynous, upon the ovary, 184.
Epilobium, 131.
Epipetalous, upon the corolla.
Epiphytes, Air Plants, 239.
Equisetaceæ, 27.   [241.
Equitant, riding astride (æstivation),
Erica, 157.
Erióphorum, 270.
Erose, eroded, as if gnawed.
Erythrònium, 30, 48, 251, 257.
Eschschóltzia, 71.
Etiolated, whitened for want of light.
Evanescent corolla, 170.
Evening Primrose, 125, 168, 173, 193.
Evergreen, 56.

Everlastings, 147.
Exalbuminous, without albumen, 186.
Excæcaria, 207.   [193, 218.
Excurrent (stem), running to the top, 107.
Exogens, 220, 229.
Exógenous structure, 108, 225.
Exserted, projecting out of or beyond.
Exstipulate, without stipules.
Extrorse (anthers), turned outward.

F.

Fagopyrum, 203.
Falcate, scythe-shaped, curved, 206.
Fascicle, a bundle, 46, 214.
Fasciculate, in a bundle, 46.
Feather-veined, see Pinni-veined.
Ferruginous, color of iron-rust.
Ferns, 20.
Fértile (flowers), producing seed, 219.
Fertilization, see Pollenization, 185, 216.
Feverfue, 147.
Fibrils, the last division of roots, 20.
Field Speedwell, 170.
Figworts, 174.
Filament, the stalk of a stamen, 32.
Filbert, 213.
Filiform, slender like a thread, 14, 260.
Filices, 26.
Fimbriate, fringed, having the border edged with slender processes, 97.
Fir, 218—Douglas Fir, 221.
Fistular, hollow, as Wheat straw.
Flabelliform, fan-shaped, 225.
Flax, Toad, 173.
Flax, New Zealand, 258.
Fleur-de-lis (Flur-de-lè), 241.
Floccous, with hairs in soft fleecy tufts.
Flora, (a) the spontaneous vegetation of a country; (b) a written description of the same.
Floral envelopes, the sepals and petals.
Florets, 138, 141.
Florets of the disk, 141.
Florets of the ray, 141.
Flowerless Plants, 16.
Flowers not made for man, 174.
Flower Region, 29.
Flowers regular, 73.
Foliaceous, leaf-like in form or texture.
Follicle, a dry, simple fruit, 1-celled, 1-valved, several-seeded, 197.
Fool's Parsley, 137.
Forked carpophore, 132.
Fork-veined. 22.
Fox Glove, 176.
Fragària, 105.
Free, not adherent to other organs.
Free Central Placenta, 163.
Fringed; see Fimbriate.
Fritillaria, 258.   [21.
Frond, an organ serving as stem and leaf,
Frutescent, shrubby.
Fruit, 33.
Fúchsia, 129.
Fugácious, soon vanishing, 18.
Fulvous, dull yellowish-brown.
Fungi, 27.

Funnel-form; see Infundibuliform.
Funículus, the seed-stem, 42, 56.
Furcate, forked.
Fusiform, spindle-shaped (root).

G.

Galánthus, 263.
Gálbanum, Gum, 137.
Galeate, the upper lip or petals arched.
Gálingale, 263. [143.
Gamopetalous, same as Monopetalous,
Garlic, 258.
Gaulthèria, 149.
Gaylussàcia, 158.
Geminate, twin, two together.
Generic characters, 179.
Geniculate, bent as the knee (*genu*).
Genus, pl. Genera, a family group, 18.
Geraniáceæ, 97.
Geránium, 40, 83.
Gerárdia, 176.
Germination, 186, 277.
Gibbous, obliquely tumid.
Glabrous, smooth, not hairy, 73, 134.
Gla-di-o-lus, 244.
Glandular, with glands, secreting organs.
Glans, a nut, as an acorn.
Glaucous, sea-green, bluish-green, usually with a bloom, or whitish powder, 34.
Globous, rounded, globular, 151.
Glumes, chaffy envelopes, 265.
Glumíferæ, the division (class) which includes the grass-like orders.
Golden Alexanders, 134.
Golden Chain, 125.
Goldenrod, 147.
Granular, composed of grains.
Grass Pink, 235.
Grasses, The, 271.
Green Dragon, 232.
Green Rose, 115.
Ground Ivy, 176.
Growth is downward, 110.
Gum Arabic, 125.
Gymnèma, the Cow Tree, 199.
Gymnosperms, with naked seeds, 220.
Gynándrous, stamens and pistils conjoined. See Column, 235.
Gynœcium, the pistils as a whole.

H.

Habit, the general aspect of a plant, 117.
Habitat, the natural locality of a plant.
Hæmanthus, 263.
Hairs, hairy, hirsute, 104.
Hastate, with the base lobes abruptly spreading, as in a halbert.
Hawthorn, 117.
Hay Fever, 32.
Hazel, 213.
Hearts-ease, 75.
Heart-wood, the duramen, 107.
Heather, 157.
Heathwort, 157.
Hedge Mustard, 185.
Heliánthus, 147.
Hellebore, 59, 64.
Hemlock, 218.

Hemp, African, 258.
Hepática, 27, 55.
Herb, a plant with an annual stem, 29.
Herb Annual, 68.
Herb Perennial, 46.
Herb Robert, 86.
Herbaceous, green and cellular, not woody.
Heronsbill, 87.
Herbárium, a collection of dried plants.
Hermáphrodite (flower), with both stamens and pistils.
Heterógamous, two sorts of flowers in the same head.
Hexandrous, with 6 stamens.
Hilum, the eye or scar of a seed.
Hip, 114.
Hirsute, hairy with rather long hairs.
Hispid, bristly with stiff hairs, 132.
Hoarhound, 182.
Hoary, frost-colored, grayish.
Holy Spirit Plant, 239.
Homógamous, head with all the flowers alike, as to stamens and pistils.
Honey, 59.
Honeysuckle, 173.
Hood, any hood-shaped organ, 197.
Hooded, see Cucullate.
Hood-leaved Violet, 74.
Hop (Húmulus), 187.
Horns, certain little projections in the Asclepias, etc., 197.
Horse Chestnut, 192.
Horse-shoe Geranium, 86.
Horsetail Rushes, 27.
Hortus siccus (hort. sic.), an herbarium.
Huckleberry (Whortleberry), 158.
Hyacinth, 35, 238.
Hyaline, transparent, or nearly so.
Hybridization, 182.
Hybrid, a cross-breed between two species.
Hypo (in composition), under; as
Hypógynous, under the ovary, or free (sc. stamens), 48.
Hypóxis, 259.
Hyssop, 182.

I.

Imbricate, imbricated, overlapping by both edges. Lily, 43, 106.
Immortélles, 147.
Inaxial root, 50.
Incised, divided deeply, as if cut.
Included, inclosed within, or shorter than.
Incumbent (embryo), 100.
Indefinite, not easily counted, 47.
Indehiscent, not opening.
Indian Cress, 95. — Cucumber, 249.
—Soap, 195.—Turnip, 231.
India Rubber, 207.
Indigenous, native of a country.
Indigo, 125.
Induplicate-valvate æstivation.
Indusium, the shield covering the fruit-dot (sorus) of a Fern.
Inferior ovary, same as adherent ovary, 133, 233.

Inflected (petal), with the point bent inward, 133.
Inflorescence, flower-arrangement, 41.
Infundibuliform corolla, the tube gradually enlarging into the limb.
Innate (anther), joined by its base to the filament, 74, 267.
Insects as pollen-bearers, 41, 48, 52, 60, 80, 82, 83, 91, 118, 128, 135, 145, 157, 163, 173, 180, 197, 201, 235, 236, 237, 240, 243, 254.
Inserted, refers to the point of junction, or apparent origin.
Integument, a coat or covering.
Internodes, 80, 85.
Introrse (anthers), turned inward, 41.
Inula, 147.
Involucre, Involucel, 132.
Involute, rolled inward.
Irregular flowers, 73.
Ipomœa, 183.
Iridaceæ, Irids, 244.
Iris, 241.
Ironwood, 213.
Isatis, 103.
Itàka, 125.
Ivy Geranium, 97.
Ivy-leaved Flax, 174.

J.

Jack-in-the-pulpit, 299.
Jacobœa, 263.
Jalap, 188.
Jatropha, 207.
Jewel Weed, 93.
Jointed, with joints, articulated, 83.
Jonquil, 261.
Juniperus, 221.

K.

Kálmia, 135.
Keeled. See Carinate.
Kino, gum, 125.
Knotweed, 200.

L.

Labiatæ, Labiate Plants, 182.
Labiate (flowers), lip-shaped, mouth-shaped, 172, 180, 182.
Labúrnum, 125.
Lady's Delight, 128.—Eardrops, 128.—Slipper, 175, 233, 240.—Thumb, 203.
Laciniate, slashed, with deep incisions.
Lactéscent, containing lac, or milk.
Lacústrine, growing in lakes.
Lambert Pine, 221.
Lámina, the blade of a leaf.
Lánceolate, lance-shaped, 14.
Lanuginous, woolly.
Lapsàna, 91.
Larkspur, 59, 64.
Latex, the turbid or milky juice of plants.
Láthyrus, 118.
Latin names of plants, 18.
Lactùca, 147.
Lateral, attached to the side (style and ovary), 106.
Laurel, The American, 155.
Laurel, Cherry, 116.

Lavándula, Lavender, 182.
Leaf-stalk, petiole, 31, etc.
Leaf-arrangement, 193.—Compound, 58.—Hues of, 199.—Modified, 120, 186, 236.—Radical, 47.—Shape depends on the venation, 190.—Use of acrid, 50.
Leaf Region, 29.
Leaf, the type of the plant, 256.
Leaflets, the pieces of a compound leaf, 53.
Leek, 258.
Legume, a simple, dry, 1-celled, 2-valved, several-seeded fruit, 118.
Leguminosæ, Leguminous Plants, 125.
Lenticular, shape of a convex lens.
Lentils, 125.
Leucòjum, 263.
Lettuce, 147.
Liber, the inner bark.
Lichens, Scale-Mosses, 16.
Ligulate (corolla), strap-shaped, 142.
Ligule, the stipules of Grasses, 271.
Liliaceæ, Lilyworts, 257.
Liliaceous flower or corolla, i. e., a 6-parted perianth.
Lilies of the Field, 62.
Lily of France, 241.
Lily of the Valley, 255.
Lily, Water, Nymphæa, 91.
Limb, the border (sc. of the flower).
Linaria, 174.
Linear, long and narrow, 14.
Linear-lanceolate, 14.
Linear-subulate, 17.
Liquorice, 125.
Liver-leaf, 54.
Liverworts, Hepaticæ, 27.
Lobed palmately,
Lobed pinnately.
Loculicidal, opening into a cell, 253.
Locust, 121.
Locústa, a spikelet of the Grasses.
Logwood, 124.
Loment, a jointed legume, 123.
Loosestrife, 165.
Lovage, 136.
Lunària, 108.
Lunate, crescent-shaped.
Lychnis, 83.
Lycopods, 27.
Lygòdium, 27.
Lyrate, or Lyrate-pinnatifid, deeply lobed in the midst (lyre-shaped?).
Lysimachia, 165. [77.

M.

Máculate, spotted or blotched.
Maidenhair, 27.
Male flowers, staminate flowers.
Mallow, 83.
Maple, 35, 93, 188, 193.
Marescent, withering while persistent.
Marigold, 147, 246.
Mast, Mastworts, 213.
Mat-grasses, 270.
Maurandia, 176.

INDEX.

Mayflower, 158.
Meadow Rue, 61.
Medéola, 250.
Medlar, 117.
Medúlla, pith; Médullary rays, 107, 130.
Membranous, thin, like a membrane.
Mentha, 182.
Merocarp, one of the carpels of a Cremocarp, 133.
Metamorphosis, a transformation.
Midrib (obsolete) the same as midvein.
Midvein, the central vein of a leaf, 21.
Milkweed, 195.
Mitriform, formed like a conical cap.
Mimosa, 124.
Modified Leaf, 120.
Moulds, 27.
Monos (in Greek compounds), one; as
Monadelphous, stamens in one set.
Monándrous, with one stamen. [91.
Monkshood, 64.
Monocárpic perennials, 263
Monocotyledonous, with one seed-lobe, 267.
Monœcious, with 2 kinds of flowers together on the same plant, 214, 231.
Monógynous, with one pistil.
Monopetalous. See Gamopetalous.
Moosewood, 192.
Morning Glory, 91, 118, 182.
Mosses, 16.
Mountain Ash. 117.
Mouse-ear Everlasting, 137.
Moving Plant, 123.
Mucronate, ending with a sharp, abrupt point (mucro), 244.
Muhlenburg, Henry, 50.
Mulberry, 116.
Mullein, 174, 177-8.
Multi (in composition), many; as
Multifid, cut half-way into many segments.
Muricate, bearing short, hard points.
Muriform, like a wall of mason-work.
Muscology, a treatise on Mosses.
Mushroom, 27.
Mustard, 102.
Mustard, Hedge, 185.
Mycélium, the first, underground growth (thallus) of the Fungi or Mushrooms.

N.

Naked receptacle, without chaff, 139.
Naked seeds, 216.
Narcissus, 261.
Napiform (root), turnip-shaped.
Nasturtion, 95.
Natant, swimming; under water.
Naturalized and Foreign Plants, 143, 176.
Nectar, the sweet secretion of flowers.
Nectarine, 116.
Nectary, an appendage secreting nectar, 47.
Nepeta, 177.
Nepenthes, 161.
Nettle, 104.
Net-veined, same as reticulate-veined, 55.

Neutral flower, one with neither stamens nor pistils, as in Hydrangea.
New Zealand Flax, 258.
Nightshade, 131.
Nipplewort, 91.
Node, nodus, a joint, 78, 85.
Nodding (flower), inclined, like the Erythronium.
Nomenclature, the rules for naming genera and species.
Normal, according to rule.
Norway Spruce, 218, 222.
Nucleus, the kernel (of ovule or seed).
Nut, same as glans.
Nutgalls, 213.
Nutgrass, 268.

O.

Oak, 35, 207.
Oats, 279.—Wild, 251.
Ob (in composition) denotes inversion; as
Obcordate, inversely heart-shaped.
Oblanceolate, inversely lance-shaped, 47.
Oblique, unequal-sided, as an Elm leaf.
Oblong, a broadly linear form.
Obovate, inversely ovate, 47.
Obsolete, past, or out of use; undeveloped, 133.
Obtuse, blunt or round at apex, 35.
Obvolute, half equitant, each leaf in the bud embracing only one margin of the other.
Sage.
Ochreæ, sheathing stipules, 200.
Ochroleucous, cream-colored, pale yellow.
Octo (in composition), eight; as
Octandrous, with 8 stamens.
Œnothera, 125.
Officinal, for sale in the shops, 171.
Officinal Speedwell, 170, 176.
Offset, a short lateral shoot.
Onagraceæ, 130.
Onion, 258.
Operculum, the lid of a Moss, 14.
Opium Poppy. 68, 70.
Opposing (petals), petals and stamens opposite, 41, 167.
Opposite (leaves), two at a node, 78.
Orbicular, circular, 152.
Orchard Grass, 274.
Orchidaceæ, 239.
Orchidaceous flower, 6-parted, 1-lipped.
Orchis, 233.
Organized, with mutually-related organs, 13.
Orontium, 232.
Orris-root, 246.
Orthótropous (ovule), erect, not bent, 214.
Osmorhiza, 131.
Osmund Fern, 24.
Osseous, bony, like the Peach stone.
Ostrya, 214.

Oval, egg-shaped with equal ends.
Ovary, 32.
Ovary, adherent and coherent, 233.
Ovary inferior, adhering to the calyx tube, 233.
Ovary superior, free from the calyx, 110.
Ovate, shape of an egg, 56.
Ovate-lanceolate, between ovate and lanceolate, 35.
Ovoid, egg-form, applied to solids, 29, 56.
Ovule erect in the cell; ascending. [cell.
Ovule suspended in the
Ovule, the young, immature seed.
Oxálides, plural of Oxalis, 89, 92.
Oxycoccus, 158.
Oxlip, 168.

P.

Pæony, 59, 64, 115.
Pales, or palæ, the inner chaff of Grasses, or of the Composites, 145, 274.
Palms, Palmaceæ, 223, 226.
Palmetto, 222.—Dwarf, 226.—Saw, 226.
Palmi-veined, or Palmate-veined, 54.
Pampas Grass, 104, 280.
Pancrátium, 263.
Panicle, a raceme compounded, 272.
Pannage, 212.
Pansy, 41, 75.
Papáver, 68.
Papaveraceæ, 67.
Papilionaceous, pa-pil'-yo-nā'-shus, 118
Pappus, the calyx of the Composites, 139.
Papyrus, 270.
Parallel-veined, 31.
Paráphyses, in the flowers of a Moss, 14.
Parénchyma, the cellular tissue, 225.
Paries, a wall; Parietal, on the wall, 66.
Parsley, 136.
Parsnip, 136.
Parthénium, 147.
Partridge-berry, 146.
Pasque Flower, 62.
Paulínia, 195.
Pea, 117, 121.
Peach, 112, 116, 120.
Peanut, 125.
Pear, 112, 116. [natifid.
Pectinate, like comb-teeth, finely pin-
Pedate, shaped like a bird's foot.
Pedicel, the divisions of a peduncle, 14, 41. [30, 40.
Peduncle, pe-dŭnk'-l. the flower-stalk,
Pelargonium, 87.
Peltate, shield-shaped, 97, 159.
Pendulous, hanging, 56.
Pennyroyal, 182.
Pentamerous, 5-parted, 173.
Pente (in composition), five; as
Pentstemon, 174, 176.
Pentandrous, with 5 stamens.
Pepo, a fruit like a melon.
Peppergrass, 103.
Peppermint, 182.
Perennial, living several years, 24, 46.
Perfect flower; see Hermaphrodite, 47.

Perfoliate, through the leaf, 251.
Peri (in composition), around; as
Perianth, the floral envelope, 31.
Pericarp, the seed-vessel. [267.
Perigynium, the perianth of a Carex,
Perigynous, inserted around the ovary, i. e. on the calyx, 106, 184, 268.
Péristome, 14.
Persian Insect Powder, 147.
Pérsica, The Peach, 201.
Persistent, remaining long in place, 33.
Personate, masked; with lips closed, 173.
Petal, the leaves of the corolla, 31.
Petalíferous, bearing petals.
Petaloid, resembling petals.
Petiolate, borne on a petiole, 54.
Pétiole, the leaf-stalk, 31.
Pétiolule, the stalk of the leaflets, 37.
Phárbitis, 187. [229.
Phenogamia, the Flowering Plants, 33,
Phœnix, 228.
Phormium, 258.
Phyllòdia, leaves without a blade.
Phyllotaxy, leaf-arrangement, 193.
Pie Plant, 203.
Pigeon-wheat Moss, 13.
Pigweed, 143.
Pilous, with erect, thin hairs.
Pine, Lambert. 221.—Long-leaved, 218.— Norfolk Island, 223. — Pitch, 217. — Prince's, 152.—Red, 217.—Weymouth, 217.—White, 214.
Pinks, 80.
Pinkworts, 83. [ions of a frond, 21.
Pinna, pl. pinnæ (wings), the divis-
Pinnate, with 4 or more lateral leaflets.—Odd pinnate, 114.
Pinnate, abruptly, with no odd leaflet.
Pinnate, interruptedly, leaflets alternately smaller.
Pinnatifid, deeply lobed in a pinnate
Pin Oak, 210. [manner, 21.
Pipsissewa, 152. [82.
Pistil, the central organ of the flower,
Pistilidia, in the flowers of a Moss, etc., 14.
Pistillate (flower), bearing pistils, 138.
Pisum, 118.
Pitch, Bergunda, etc., 223.—Pine, 215.
Pitcher Plant, 158.
Pitted tissue, 221.
Placenta, pl. Placentæ, the cellular part of the carpel which bears the ovules.
Placentæ central, 66.
Placentæ free central, 162.
Placentæ parietal, on the wall, 66.
Plan of a flower, 32.
Plantain, 82, 143. 177.
Plants, Carnivorous. 161.—Flowering, 33. —Flowerless, 16.—Food of, 107.—Sleep of, 29, 91.—Tropical, 146.

INDEX. 315

Pleurisy Root, 199.
Plicate, folded like a fan, 183.
Plumous, like a plume; feathery.
Plumule (a little plume), 111.
Poa, 271.
Poet's Narcissus, 262.
Poinciàna, 125.
Pollen, abundance of, 32.
Pollenization, curious facts in, 41, 60, 74, 77, 80, 82, 85, 91, 118, 128, 135, 145, 157, 163, 173, 180, 197, 201, 207, 235, 236, 237, 240, 243.
Pollinia, masses of pollen, 197.
Poly (in composition), many; as
Polyándrous, with many stamens,
Polyánthus 168, 261, 263. [47.
Polygonaceæ, 203.
Polygonum, 200.
Polypetalous, with the petals free and distinct, 142.
Polypod Fern, 20.
Polytrichum, 15.
Pome, a fruit like an apple, 108.
Poor-man's-weather-glass, 168.
Poppy, 68.
Poppy Bee, 70.
Poppyworts, 67, 70.
Portulaca, 43, 91.
Portulacaceæ, 43.
Posterior, next to the axis.
Potato, Sweet, 187.
Potentilla, 112.
Precocious, flowering before leafing.
Premorse, ending abruptly, 246.
Prickles, distinguished from thorns, 114.
Pride-of-Ohio, 161.
Primrose, 125, 168.
Primulaceæ, Primworts, 161, 168.
Prince's Pine, 152.
Prismatic, shaped like a prism, 3, 4, or many-sided.
Procumbent (stem), lying prostrate.
Produced, unusually extended.
Proterándrous, 82.
Proliferous, reproducing, as cymes from the midst of a cyme, flowers from the midst of a flower.
Prothallus, 22.
Provinces, 33, 43.
Prunus, 116.
Pteris, 27.
Pubescent, downy with short, soft hairs.
Puberulent, minutely downy.
Pulsatilla, 62. [needle.
Punctate, dotted, as if punctured with a
Purple-fringed Orchis, 238.
Purslanes, 43.
Pyriform, of the form of a pear.
Pyrus, 112, 117.
Pyrola, 150.
Pyxis, a pericarp with a lid, 43, 169.

Q.

Quadrángular, four-angled.
Quality, the external traits, affecting the senses, 260.
Quamash, 258.
Quercus, 208.

Quince, 117.
Quinate, growing in fives.
Quincuncial æstivation, 42, 106.
Quinque (in composition), five.

R.

Raceme, flowers arranged as in Currant, 41, 201.
Rachis, the axis of an inflorescence, etc., 24, 41.
Radiant Head, flowers all ligulate, 144.
Radiate head, the outer row of flowers ligulate, 145.
Radical, springing from the root, 47, 54.
Radicle, the root end of the embryo.
Radicle accumbent, 108. See Accumbent.
Radicle incumbent, 100. See Incumbent.
Radish, 108.
Ramial, of a branch (ramus).
Ranunculaceæ, 59, 62.
Ranúnculus, 48.
Ranstead, 174.
Rape-seed oil, 108.
Rays (of the Composites), 142.
Rays of an Umbel, 132.
Receptacle, where the florets of a Composite stand, 139.
Receptacle chaffy, 145.
Receptacle naked, 139.
Reclined, the leaf in bud bent over forward.
Recurved, bent (not rolled) backward, 31.
Red Maple, 192.
Red Oak, 210.
Reduplicate-valvate æstivation, the valves with recurved edges.
Reflexed, bent back excessively, 52.
Regma, the fruit of the Geranium, 85.
Regular, like parts similar, 73.
Reniform, kidney-shaped, 73, 177.
Repand toothed, 126.
Reproductive organs, 29.
Resupinate, reversed; upside down, 72.
Reticulate, netted, 47.
Retuse, the apex broadly indented.
Revolute, rolled backward.
Rheum, 203.
Rhizoma, Rhizome, 20.
Rhombic, of the shape of a rhomb.
Rhododendron, 158.
Rhubarb, 203.
Ribs, ridges on the fruit of the Umbelworts, 135.
Richardia, 232.
Ricinus, 206.
Ringent (corolla), the throat open.
Robinia, 121.
Robin's Plantain, 140.
Rock Maple, 188.

316 INDEX.

Root, the base of the plant, 13, 107.
Root, axial and inaxial, 50.
Rootlets, divisions of the root, 13.
Root-stock, the rhizome.
Rosa, Rose, 112.
Rosaceæ, Roseworts, 115.
Rostrate, beaked, with a beak.
Rosaceous corolla, rose-like, viz., with 5, regular, quincuncial petals.
Rotate (corolla), monopetalous, wheel-shaped, 171.
Rubus, 116.
Rudiment, the beginning of a thing.
Rugous, wrinkled, 206.
Rue Anemone, 60.
Rue, Meadow, 60.
Rumex, 203.
Runner, a prostrate branch, 104.
Runcinate, hooked backward, 143.
Rush, 264.
Russellia, 176.
Rye, 279.

S.

Sabal Palmetto, 223.
Safflower, 246.
Saffron, 147, 246.
Sage, 180.
Sagittate, arrow-shaped, 98, 260.
Sago Palm, Sagus, 228.
Salsify, 147.
Salver-shaped corolla, a flat border with a slender tube, like Phlox. [fruit, 191.
Samara, a simple, winged
Sanguinària, 64.
Sapindàceæ, Sapindus, 195.
Saponaria, 83.
Sarracénia, 158.
Scabrous, rough, 142, 275.
Scales of the involucre, 138
Scale-mosses, 27.
Scámmony, 188.
Scarious, dry and translucent, 138, 244.
Scape, a radical flower-stalk, 56.
Scape-like, stem with diminished leaves.
Scilla, 258.
Scientific Names, Use of, 18.
Scorpoid raceme, rolled inward, and unrolling as it blossoms.
Scrophulariaceæ, 174.
Sea-weeds, 27.
Secund, turned to one side, 256.
Sections, cuttings, 184.
Sedges, 263.
Seed, Importance of, 33.
Seed, Composition of, 186.
Seed, Life of, 185.
Self-heal, 179.
Semi (in composition), half ; as
Semicordate, half cordate.
Senegal Gum, 125.
Senna, 125.
Sensitiveness, 121.
Sensitive Plant, 123.
Sensivèra, Hemp, 258.
Sépals, the leaves of the calyx, 31,

Septicidal, opening between the cells.
Septif'ragal, valves breaking away from the partitions, which remain in place, 185.
Septum, a partition.
Sequoya, 221.
Sericeous, silky.
Serótinous, occurring late in the season.
Serrate, saw-edged, 104.
Serrulate, finely saw-toothed, 14, 171.
Sessile (sitting), having no petiole or foot-stalk, 37.
Seta, plu. setæ (bristles), perianth of the Sedges.
Shadberry, 116. [270.
Sheath, the petiole of the Grasses, 271.
Sheathing petiole, 132.
Sheep Poison, 157.
Sheep Sorrel, 203. [274.
Shepherd's Purse, 98, 177,
Shooting Star, 161.
Showy Orchis, 234.
Shrub, a small (6-20 ft.), woody plant.
Silene, 83.
Silicle, Silique (all-ēke), 100, 101.
Silk Grass, 195.
Silver grain, the medullary rays, 107.
Simple, of one piece, not compound, 18.
Sinuate, margin with rounded lobes and sinuses.
Siphònia, 207.
Sisymbrium, 185.
Silver-leaved Maple, 192.
Sleep of plants, 31, 91, 92. [128.
Slips, cuttings which grow when severed,
Snap Dragon, 173, 175.
Snow Drop, 263.
Snow Flake, 263.
Soapberry, 195.
Soapworts, 83, 195.
Social Flowers, 135.
Solidago, 147.
Solitary inflorescence, 31.
Solvent trunk or axis, 107.
Sorrelworts, 203.
Sorosis, fruit compounded of an inflorescence, as Pineapple.
Sorus, pl. sori, fruit-dots of Ferns, 22.
Spadix, a spike with a fleshy rachis.
Spanish Chestnut, 213. [231.
Spāthe, the bract (colored) sheathing a spadix, 231.
Spatulate, form of a surgeon's spatula.
Spearmint, 182.
Species, 18.
Specific Characters, 179.
Speedwell, 170.
Spike, the flowers sessile on the rachis.
Spikelets, the peculiar clusters in the Grasses, 272.
Spike, Oil of. 182.
Spines, woody thorns, 121.

INDEX. 317

Spiral arrangement of leaves, 193.
Spiral vessels or cells, 189, 225.
Spirèa, 117.
Spores, the seed of the Cryptogams, 14, 22.
Sporangia, the vessels containing spores, [22.
Spotted Chimáphila, 154.
Sprekéllia, 263.
Spring Beauty, 89.
Spruce, 218.
Spur, a floral appendage, or nectary, 97, 235. Columbine.
Spurge, Spotted, 206.
Spurgeworts, 206.
Squills, 258.
Squirting Cucumber, 95.
Stamens, 32.
Staminate flower, 137.
Standard, or banner, 118.
Stapelia, 199.
Staphylèa, 195.
Star Grass, 258.
Star of Bethlehem, 258.
Stellaria, 80.
Stem, the ascending axis.
Sterile flower, not fruitful, 137, 191.
Stigma discoid—stellate.
Stigma plumous, as in Grasses.
Stigma, 32.
Stigmatic, partaking of the stigma.
Stings, hollow, poisonous hairs, 104.
Stipe, the stalk of the ovary, 21.
Stipels, the stipules of the leaflets, 117.
Stipitate, on a stipe.
Stipules, small leaves at base of the petiole, always in pairs, 75.
Stolon, a runner.
Stoloniferous, producing stolons, 137.
Storksbill, 87.
Stoma, mouth (of a sporange), 14.
Stomata, mouths in the cuticle of leaves.
Strawberry, 104.
Strict, straight and erect.
Striped Maple, 192.
Strobile, fruit of the Pines ; a cone.
Struggle for existence, 146.
Style, the middle part of the pistil, 32.
Sub (in composition), under ; in a less degree.
Subkingdoms, 22.
Subulate, awl-shaped, 17.
Succulent, very juicy and cellular.
Suffrùticous, partly shrubby (*frutex*, a shrub). 154.
Superior (ovary), ovary free.
Superior (calyx), calyx adherent.
Sugar Maple, 188.
Sundew, 104, 161.
Sunflower, 145, 147.
Supérvolute æstivation, 183.
Suppression, 194, 209.
Suspended ovule, growing from the top of the cell.
Sutural (dehiscence), opening at the sutures.
Suture (sûte-yur), 66, 118.

Swamp Maple, 192.—Milkweed, 195.
Sweet Alyssum, 103.—Flag, 233.—Pea, 117.—Vernal Grass, 276.—Violet, 75.—William, 83.
Symmetrical, of the same number.
Syn (in composition), together.
Syngenècious, stamens united by their anthers, as in the Composites.
Synonym, 103.
Sysirinchium, 244.

T.

Tagètes, 147.
Tamarind, 125.
Tannic acid, 223.
Tapioca, 207.
Tap root. See Axial root.
Taráxacum, 144.
Tawny, fulvous, dull yellowish brown.
Tea-berry, 147. [seed," 111.
Tegmen, "inner layer of the coating of a
Tendril, an appendage for climbing, 119.
Teratology, 82.
Terete (stem), evenly rounded, cylindric, 18.
Terminal, placed at the summit or apex, 14. [246.
Ternate (leaves, or leaflets), in threes, 47.
Testa, the outer coat of a seed, 33.
Tet-ra-dyn-a-mous, 4 stamens longer than the other 2, 100.
Thalictrum, 59.
Thallus, the cellular body of a Lichen, etc., bearing the fructification.
Thimble-berry, 116.
Thistle, 147.
Thorns. See Spines, 114, 121.
Throat, orifice of a monopetalous corolla.
Thyrse, a dense panicle, as in Lilac, Horse Tigridia, 246. [Chestnut.
Toad Flax, 173.
Tolu gum, 125. [hairs, 110.
Toméntous, with short, dense, woolly
Tonga Bean, 125.
Toothroot Cress, 101.
Top-shaped, inversely conical.
Torrey, Dr. John, 265.
Torus, the basis of a flower, 31.
Trág'acanth, 125.
Tragopògon, 147.
Tree, 107.
Tri (in composition), three ; as
Triandrous, having 3 stamens.
Tric'olor (three-colored), 75.
Trientàlis, 165.
Trifid, cut deeply in 3 parts.
Trifoliolate, with 3 leaflets.
Trillium, 246.
Trilliaceæ, 251.
Tríl-o-bate, having 3 lobes.
Trím-e-rous, 3-parted.
Tripínnate, thrice pinnate.
Triquetrous, three-angled, equitant æstivation, 260.
Tri-ter-nate, thrice ternate, 58.

318 INDEX.

Tri-tònia, 246.
Tropæolum, 97.
Tropical vegetation, Luxuriance of, 146.
Trophyworts, 97.
Truncate, cut square off, 180.
Trunk, the stem of a tree, 107.
Tryma, a bony fruit, like the Hickory nut.
Tuber, a thickened, underground stem, as a potato.
Tuberculate, covered with warts (tubercles).
Tùberose (Tu-ber-ose), 263.
Tubular corolla, 143.
Tule, 270.
Tulip, 35, 258.
Tumid, swelled or inflated.
Tunicated (bulb), with the layers entire, [34.
Turnip, 103.
Turpentine, 223.

U.

Umbel, Umbellet, 59, 132.
Umbellate, bearing umbels.
Umbelliferæ, 136, 184.
Unarmed, without stings, thorns, etc.
Undershrub, a low shrub.
Undulate, wavy.
Unguiculate (petal), having a claw (or petiole), 97.
Uni (in composition), one; as
Uni-valved, with one valve.
Unsymmetrical, 100.
Urceolate, urn-shaped, 149.
Utricle, a fruit with one seed loose in the thin shell.
Uva-ursi, 158.
Uvulària, 252.

V.

Vaccínium, 148, 158.
Vaginate (petiole), sheathing.
Valvate, opening by or like valves, 43.
Valvate æstivation, the pieces meeting edge to edge.
Valves, the pieces of a capsule, legume, etc.
Vanilla, 240.
Varieties, 36.
Vascular tissue, composed of vessels and tubes rather than cells ; as the Flowering Plants generally.
Vaulted, arched above, as the upper lip of some Labiates.
Vegetative Organs, 29.
Veins, Veinlets, Veinulets, 21. [22.
Venation, the arrangement of the veins,
Ventral, in front, opposite the axis.
Venus's Fly Trap, 160.

Vernal, in or pertaining to the Spring.
Vernation, arrangement of the leaves in the bud, 24.
Veronica, 170.
Vérsatile (anther), 85, 128.
Vertical, parallel with the axis, or up and down.
Verticils, whorls, 179, 198.
Ver-tic'-il-late, arranged in verticils, 154, 165.
Véspertine, appearing in the evening.
Vexillary (æstivation), like that of the Pea, 118.
Vexíllum, the banner, 118.
Villous, with long, weak hairs, 102.
Vine, a weak, slender stem, usually climbing.
Violaceæ, Violetworts, 76.
Violet, Viola, 72, 115.
Viscid, viscous, sticky.
Vitality of seeds, 185.
Vitta, pl. Vittæ, the oil-tubes in the fruit of the Umbelworts, 135.

W.

Wake Robin, 249.
Wall Flower, 103.
Water Lily, 91.
Watsònia, 246.
Wax Plant, 199.
Wedge-shaped, tapering to the base, cuneiform.
Weymouth Pine, 217.
Whistle Wood, 192.
White Maple, 192.
White Oak, 207.
White Pine, 214.
Whorl, a circle of similar organs, 154.
Whorled, see Verticillate.
Whortleberry, 158.
Wild Oats, 251.
Willow, 104.
Willow Herb, 131.
Wind Fertilization, 207, 216.
Wing-margined, 174.
Wintergreen, 147.
Wintergreen, Chick, 164.
Wistaria, 117, 125.
Wood, the structure of, 107.
Wood Anémone, 60.
Wood Sorrel, 89.
Wormwood, 147.

Y.

Yellow Dock, 50.
Yucca, 258.

Z.

Zauschnèria, 131.
Zínnia, 147.

THE NATIONAL SERIES OF STANDARD SCHOOL-BOOKS.

GEOGRAPHY.

MONTEITH'S SYSTEM.

TWO-BOOK SERIES. INDEPENDENT COURSE.

Elementary Geography.
Comprehensive Geography (with 103 maps).

☞ These volumes are not revisions of old works, not an addition to any series, but are entirely new productions, — each by itself complete, independent, comprehensive, yet simple, brief, cheap, and popular; or, taken together, the most admirable "series" ever offered for a common-school course. They present the following features, skilfully interwoven, the student learning all about one country at a time. Always revised to date of printing.

LOCAL GEOGRAPHY. — Or, the Use of Maps. Important features of the maps are the coloring of States as objects, and the ingenious system for laying down a much larger number of names for reference than are found on any other maps of same size, and without crowding.

PHYSICAL GEOGRAPHY. — Or, the Natural Features of the Earth; illustrated by the original and striking RELIEF MAPS, being bird's-eye views or photographic pictures of the earth's surface.

DESCRIPTIVE GEOGRAPHY. — Including the Physical; with some account of Governments and Races, Animals, &c.

HISTORICAL GEOGRAPHY. — Or, a brief summary of the salient points of history, explaining the present distribution of nations, origin of geographical names, &c.

MATHEMATICAL GEOGRAPHY. — Including Astronomical, which describes the Earth's position and character among planets; also the Zones, Parallels, &c.

COMPARATIVE GEOGRAPHY. — Or, a system of analogy, connecting new lessons with the previous ones. Comparative sizes and latitudes are shown on the margin of each map, and all countries are measured in the "frame of Kansas."

TOPICAL GEOGRAPHY. — Consisting of questions for review, and testing the student's general and specific knowledge of the subject, with suggestions for geographical compositions.

ANCIENT GEOGRAPHY. — A section devoted to this subject, with maps, will be appreciated by teachers. It is seldom taught in our common schools, because it has heretofore required the purchase of a separate book.

GRAPHIC GEOGRAPHY, or Map-Drawing by Allen's "Unit of Measurement" system (now almost universally recognized as without a rival), is introduced throughout the lessons, and not as an appendix.

CONSTRUCTIVE GEOGRAPHY. — Or, Globe-Making. With each book a set of map segments is furnished, with which each student may make his own globe by following the directions given.

RAILROAD GEOGRAPHY. — With a grand commercial map of the United States, illustrating steamer and railroad routes of travel in the United States, submarine telegraph lines, &c. Also a "Practical Tour in Europe."

MONTEITH AND McNALLY'S SYSTEM.

THREE AND FIVE BOOKS. NATIONAL COURSE.

Monteith's First Lessons in Geography.
Monteith's New Manual of Geography.
McNally's System of Geography.

The new edition of McNally's Geography is now ready, rewritten throughout by James Monteith and S. C. Frost. In its new dress, printed from new type, and illustrated with 100 new engravings, it is the latest, most attractive, as well as the most thoroughly practical book on geography extant.

THE NATIONAL SERIES OF STANDARD SCHOOL-BOOKS.

BARNES'S NEW MATHEMATICS.

In this series JOSEPH FICKLIN, Ph. D., Professor of Mathematics and Astronomy in the University of Missouri, has combined all the best and latest results of practical and experimental teaching of arithmetic with the assistance of many distinguished mathematical authors.

Barnes's Elementary Arithmetic.
Barnes's National Arithmetic.

These two works constitute a *complete arithmetical course in two books.*

They meet the demand for text-books that will help students to acquire the greatest amount of useful and practical knowledge of Arithmetic by the smallest expenditure of *time, labor,* and *money.* Nearly every topic in Written Arithmetic is introduced, and its principles illustrated, by exercises in *Oral* Arithmetic. The free use of Equations; the concise method of combining and treating Properties of Numbers; the treatment of Multiplication and Division of Fractions in *two* cases, and then reduced to *one*; Cancellation by the use of the vertical line, especially in Fractions, Interest, and Proportion; the brief, simple, and greatly superior method of working Partial Payments by the "Time Table" and Cancellation; the substitution of formulas to a great extent for rules; the full and practical treatment of the Metric System, &c., indicate their completeness. A *variety* of methods and processes for the *same topic*, which deprive the pupil of the great benefit of doing a part of the *thinking* and *labor* for himself, have been discarded. The statement of principles, definitions, rules, &c., is brief and simple. The illustrations and methods are explicit, direct, and practical. The great number and variety of Examples embody the actual business of the day. The very large amount of matter condensed in so small a compass has been accomplished by economizing every line of space, by rejecting superfluous matter and obsolete terms, and by avoiding the *repetition* of analyses, explanations, and operations in the advanced topics which have been used in the more elementary parts of these books.

AUXILIARIES.

For use in district schools, and for supplying a text-book in advanced work for classes having finished the course as given in the ordinary Practical Arithmetics, the National Arithmetic has been divided and bound separately, as follows:—

Barnes's Practical Arithmetic.
Barnes's Advanced Arithmetic.

In many schools there are classes that for various reasons never reach beyond Percentage. It is just such cases where *Barnes's Practical Arithmetic* will answer a good purpose, at a *price to the pupil* much less than to buy the complete book. On the other hand, classes having finished the ordinary Practical Arithmetic can proceed with the higher course by using *Barnes's Advanced Arithmetic.*

For primary schools requiring simply a table book, and the earliest rudiments forcibly presented through object-teaching and copious illustrations, we have prepared

Barnes's First Lessons in Arithmetic,

which begins with the most elementary notions of numbers, and proceeds, by simple steps, to develop all the fundamental principles of Arithmetic.

Barnes's Elements of Algebra.

This work, as its title indicates, is elementary in its character and suitable for use, (1) in such public schools as give instruction in the Elements of Algebra; (2) in institutions of learning whose courses of study do not include Higher Algebra; (3) in schools whose object is to prepare students for entrance into our colleges and universities. This book will also meet the wants of students of Physics who require some knowledge of

THE NATIONAL SERIES OF STANDARD SCHOOL-BOOKS.

Algebra. The student's progress in Algebra depends very largely upon the proper treatment of the four *Fundamental Operations*. The terms *Addition, Subtraction, Multiplication,* and *Division* in Algebra have a wider meaning than in Arithmetic, and these operations have been so defined as to *include* their arithmetical meaning; so that the beginner is simply called upon to *enlarge* his views of those fundamental operations. Much attention has been given to the explanation of the negative sign, in order to remove the well-known difficulties in the use and interpretation of that sign. Special attention is here called to "A Short Method of Removing Symbols of Aggregation," Art. 76. On account of their importance, the subjects of *Factoring, Greatest Common Divisor,* and *Least Common Multiple* have been treated at greater length than is usual in elementary works. In the treatment of *Fractions*, a method is used which is quite simple, and, at the same time, more general than that usually employed. In connection with *Radical Quantities* the roots are expressed by fractional exponents, for the principles and rules applicable to integral exponents may then be used without modification. The *Equation* is made the chief subject of thought in this work. It is defined near the beginning, and used extensively in every chapter. In addition to this, four chapters are devoted exclusively to the subject of *Equations*. All *Proportions* are equations, and in their treatment as such all the difficulty commonly connected with the subject of Proportion disappears. The chapter on Logarithms will doubtless be acceptable to many teachers who do not require the student to master Higher Algebra before entering upon the study of Trigonometry.

HIGHER MATHEMATICS.

Peck's Manual of Algebra.
Bringing the methods of Bourdon within the range of the Academic Course.

Peck's Manual of Geometry.
By a method purely practical, and unembarrassed by the details which rather confuse than simplify science.

Peck's Practical Calculus.

Peck's Analytical Geometry.

Peck's Elementary Mechanics.

Peck's Mechanics, with Calculus.
The briefest treatises on these subjects now published. Adopted by the great Universities: Yale, Harvard, Columbia, Princeton, Cornell, &c.

Macnie's Algebraical Equations.
Serving as a complement to the more advanced treatises on Algebra, giving special attention to the analysis and solution of equations with numerical coefficients.

Church's Elements of Calculus.

Church's Analytical Geometry.

Church's Descriptive Geometry. With plates. 2 vols.
These volumes constitute the "West Point Course" in their several departments. Prof. Church was long the eminent professor of mathematics at West Point Military Academy, and his works are standard in all the leading colleges.

Courtenay's Elements of Calculus.
A standard work of the very highest grade, presenting the most elaborate attainable survey of the subject.

Hackley's Trigonometry.
With applications to Navigation and Surveying, Nautical and Practical Geometry, and Geodesy.

THE NATIONAL SERIES OF STANDARD SCHOOL-BOOKS.

DR. STEELE'S ONE-TERM SERIES, IN ALL THE SCIENCES.

Steele's 14-Weeks Course in Chemistry.
Steele's 14-Weeks Course in Astronomy.
Steele's 14-Weeks Course in Physics.
Steele's 14-Weeks Course in Geology.
Steele's 14-Weeks Course in Physiology.
Steele's 14-Weeks Course in Zoölogy.
Steele's 14-Weeks Course in Botany.

Our text-books in these studies are, as a general thing, dull and uninteresting. They contain from 400 to 600 pages of dry facts and unconnected details. They abound in that which the student cannot learn, much less remember. The pupil commences the study, is confused by the fine print and coarse print, and neither knowing exactly what to learn nor what to hasten over, is crowded through the single term generally assigned to each branch, and frequently comes to the close without a definite and exact idea of a single scientific principle.

Steele's "Fourteen-Weeks Courses" contain only that which every well-informed person should know, while all that which concerns only the professional scientist is omitted. The language is clear, simple, and interesting, and the illustrations bring the subject within the range of home life and daily experience. They give such of the general principles and the prominent facts as a pupil can make familiar as household words within a single term. The type is large and open; there is no fine print to annoy; the cuts are copies of genuine experiments or natural phenomena, and are of fine execution.

In fine, by a system of condensation peculiarly his own, the author reduces each branch to the limits of a single term of study, while sacrificing nothing that is essential, and nothing that is usually retained from the study of the larger manuals in common use. Thus the student has rare opportunity to *economize his time*, or rather to employ that which he has to the best advantage.

A notable feature is the author's charming "style," fortified by an enthusiasm over his subject in which the student will not fail to partake. Believing that Natural Science is full of fascination, he has moulded it into a form that attracts the attention and kindles the enthusiasm of the pupil.

The recent editions contain the author's "Practical Questions" on a plan never before attempted in scientific text-books. These are questions as to the nature and cause of common phenomena, and are not directly answered in the text, the design being to test and promote an intelligent use of the student's knowledge of the foregoing principles.

Steele's Key to all His Works.

This work is mainly composed of answers to the Practical Questions, and solutions of the problems, in the author's celebrated "Fourteen-Weeks Courses" in the several sciences, with many hints to teachers, minor tables, &c. Should be on every teacher's desk.

Prof. J. Dorman Steele is an indefatigable student, as well as author, and his books have reached a fabulous circulation. It is safe to say of his books that they have accomplished more tangible and better results in the class-room than any other ever offered to American schools, and have been translated into more languages for foreign schools. They are even produced in raised type for the blind.

THE NATIONAL SERIES OF STANDARD SCHOOL-BOOKS.

NATURAL SCIENCE — *Continued.*

BOTANY.

Wood's Object-Lessons in Botany.
Wood's American Botanist and Florist.
Wood's New Class-Book of Botany.
The standard text-books of the United States in this department. In style they are simple, popular, and lively; in arrangement, easy and natural; in description, graphic and scientific. The Tables for Analysis are reduced to a perfect system. They include the flora of the whole United States east of the Rocky Mountains, and are well adapted to the regions west.

Wood's Descriptive Botany.
A complete flora of all plants growing east of the Mississippi River.

Wood's Illustrated Plant Record.
A simple form of blanks for recording observations in the field.

Wood's Botanical Apparatus.
A portable trunk, containing drying press, knife, trowel, microscope, and tweezers, and a copy of Wood's "Plant Record," — the collector's complete outfit.

Willis's Flora of New Jersey.
The most useful book of reference ever published for collectors in all parts of the country. It contains also a Botanical Directory, with addresses of living American botanists.

Young's Familiar Lessons in Botany.
Combining simplicity of diction with some degree of technical and scientific knowledge, for intermediate classes. Specially adapted for the Southwest.

Wood & Steele's Botany.
See page 33.

AGRICULTURE.

Pendleton's Scientific Agriculture.
A text-book for colleges and schools; treats of the following topics: Anatomy and Physiology of Plants; Agricultural Meteorology; Soils as related to Physics; Chemistry of the Atmosphere; of Plants; of Soils; Fertilizers and Natural Manures; Animal Nutrition, &c. By E. M. Pendleton, M. D., Professor of Agriculture in the University of Georgia.

From PRESIDENT A. D. WHITE, *Cornell University.*

"*Dear Sir:* I have examined your 'Text-book of Agricultural Science,' and it seems to me excellent in view of the purpose it is intended to serve. Many of your chapters interested me especially, and all parts of the work seem to combine scientific instruction with practical information in proportions dictated by sound common sense."

From PRESIDENT ROBINSON, *of Brown University.*

"It is scientific in method as well as in matter, comprehensive in plan, natural and logical in order, compact and lucid in its statements, and must be useful both as a text-book in agricultural colleges, and as a hand-book for intelligent planters and farmers."

THE NATIONAL SERIES OF STANDARD SCHOOL-BOOKS.

MODERN LANGUAGES.

A COMPLETE COURSE IN THE GERMAN.

By James H. Worman, A.M., Professor of Modern Languages in the Adelphi Academy, Brooklyn, L. I.

Worman's First German Book.
Worman's Second German Book.
Worman's Elementary German Grammar.
Worman's Complete German Grammar.

These volumes are designed for intermediate and advanced classes respectively.

Though following the same general method with "Otto" (that of "Gaspey"), our author differs essentially in its application. He is more practical, more systematic more accurate, and besides introduces a number of invaluable features which have never before been combined in a German grammar.

Among other things, it may be claimed for Professor Worman that he has been *the first* to introduce, in an American text-book for learning German, a system of analogy and comparison with other languages. Our best teachers are also enthusiastic about his methods of inculcating the art of speaking, of understanding the spoken language, of correct pronunciation; the sensible and convenient original classification of nouns (in four declensions), and of irregular verbs, also deserves much praise. We also note the use of heavy type to indicate etymological changes in the paradigms and, in the exercises, the parts which specially illustrate preceding rules.

Worman's Elementary German Reader.
Worman's Collegiate German Reader.

The finest and most judicious compilation of classical and standard German literature. These works embrace, progressively arranged, selections from the masterpieces of Goethe, Schiller, Korner, Seume, Uhland, Freiligrath, Heine, Schlegel, Holty, Lenau, Wieland, Herder, Lessing, Kant, Fichte, Schelling, Winkelmann, Humboldt, Ranke, Raumer, Menzel, Gervinus, &c., and contain complete Goethe's "Iphigenie," Schiller's "Jungfrau;" also, for instruction in modern conversational German, Benedix's "Eigensinn."

There are, besides, biographical sketches of each author contributing, notes, explanatory and philological (after the text), grammatical references to all leading grammars, as well as the editor's own, and an adequate Vocabulary.

Worman's German Echo.
Worman's German Copy-Books, 3 Numbers.

On the same plan as the most approved systems for English penmanship, with progressive copies.

CHAUTAUQUA SERIES.

First and Second Books in German.

By the natural or Pestalozzian System, for teaching the language without the help of the Learner's Vernacular. By James H. Worman, A. M.

These books belong to the new Chautauqua German Language Series, and are intended for beginners learning to *speak* German. The peculiar features of its method are:—

1. It teaches the language by direct appeal to illustrations of the objects referred to, and does not allow the student to guess what is said. He speaks from the first hour *understandingly* and *accurately.* Therefore,

2. Grammar is taught both analytically and synthetically throughout the course. The beginning is made with the auxiliaries of tense and mood, because their kinship with the English makes them easily intelligible; then follow the declensions of nouns, articles, and other parts of speech, always systematically arranged. It is easy to confuse the pupil by giving him one person or one case at a time. This pernicious practice is discarded. Books that beget unsystematic habits of thought are worse than worthless.

www.ingramcontent.com/pod-product-compliance
Lightning Source LLC
Chambersburg PA
CBHW030736230426
43667CB00007B/730